D0893933

RADIO INTERFACE SYSTEM PLANNING FOR GSM/GPRS/UMTS

RADIO INTERFACE SYSTEM PLANNING FOR GSM/GPRS/UMTS

By

Jukka Lempiäinen

Matti Manninen

KLUWER ACADEMIC PUBLISHERS
BOSTON / DORDRECHT / LONDON

A C.I.P. Catalogue record for this book is available from the Library of Congress.

ISBN 0-7923-7516-5

Published by Kluwer Academic Publishers,
P.O. Box 17, 3300 AA Dordrecht, The Netherlands.

Sold and distributed in North, Central and South America
by Kluwer Academic Publishers,
101 Philip Drive, Norwell, MA 02061, U.S.A.

In all other countries, sold and distributed
by Kluwer Academic Publishers,
P.O. Box 322, 3300 AH Dordrecht, The Netherlands.

Printed on acid-free paper

Printed in the Netherlands.

To my wife Tia,
to my parents Irene and Keijo,
to my brother Jarmo in memoriam.

J. Lempiäinen

To my wife Marika.

M. Manninen

TABLE OF CONTENTS

PREFACE

There are plenty of books written for academic courses about the areas of electromagnetics, radio technology or radio communications. In these books all the basic and more of the advanced mathematical and technical engineering principles about the theory of electromagnetics—reflections, diffractions, scattering, coupling-antennas and radio propagation—are explained. There are also many books concerned only radio communications and especially mobile communication systems. These books explain theoretical aspects of mobile communication systems in more detail but do not focus on the connections between different technical details. System level engineering is thus not addressed because it is typically thought only to be a practical issue. However, this system level understanding is a critical issue in today's mobile communications systems in order to achieve cost efficient and high quality radio engineering in mobile systems' radio interface. It is not enough to understand just technical details because these are connected to other technical aspects and these can all be combined in many different ways.

These various technical combinations within this mobile communications branch can be demotivating, particularly to newcomers—new RF engineers—who already know all the basics of radio technology because they have read all the basic books. Moreover, new engineers are often confused with all the different solutions for the different problems when they start learning the mobile communications principles and they can easily feel frustration because system level planning is a new and quite poorly documented topic. Hence, a simple and clear handbook about the radio planning principles and their connections in mobile communications systems is required to explain this system level planning.

This book is a part of the solution to understanding detailed technical radio engineering principles for mobile communication systems where all the details have a straight connection with each other. This book tries to bind very detailed pieces of technical information to the system level of thinking and thus to show the correct path at the junction. Mobile communication is full of these junctures because almost all the products are new, and technical development work is only just started. This handbook has been produced mainly with examples from the Global System for Mobile Telecommunication (GSM) because it is one of the most "global" standards (implemented over 80 countries). There are also many technical topics to be developed for the GSM during the next decade when GSM as a system is combined together with the Universal Mobile Telecommunication System (UMTS) that will be the next GSM compatible standard. The GSM examples in this book, and most of the technologies utilized in the GSM, can also thus be applied in the UMTS. Clear technical differences in the radio system planning between GSM and UMTS system are be explained and discussed in the concluding chapters 10 and 11 but otherwise there is no need to explain the variations between GSM and UMTS in technical detail here because this is covered in several other books. The main purpose of this handbook is to explain the radio interface system planning in the GSM (including GPRS) and to give the main instructions to continue the same work in the UMTS.

The content of this book is based on the radio system planning process that contains different phases that are explained in Chapters 3–9. Before these detailed planning phases Chapter 1 gives an introduction to the radio propagation environment in order to convey the influence of the environment on the radio system planning principles. The whole radio system planning process is also explained separately in Chapter 2 and Chapters 3–9 following the content of Chapter 2 at a detailed level. Radio system planning starts from configuration planning (Chapter 3) and continues with coverage planning (Chapters 4 and 5), capacity planning (Chapter 6), frequency planning (Chapter 7) and optimisation and monitoring (Chapters 8 and 9). Chapter 3 considers the definitions of the power budget calculations; the coverage planning done in Chapters 4 and 5 is based on the defined power budget from Chapter 3. Chapter 6 considers the capacity of the radio interface and Chapter 7 contains the frequency planning aspects. Chapters 8 and 9 give instructions on observing the performance of the planned radio system/network. Finally, Chapters 10 and 11 explain the detailed radio planning differences of the speech orientated GSM radio network and the packet transferred data GPRS/EDGE and the third generation UMTS radio networks.

Jukka Lempiäinen
Matti Manninen

Helsinki, July 2001

Chapter 1

INTRODUCTION— RADIO PROPAGATION ENVIRONMENT

1. INTRODUCTION— RADIO PROPAGATION ENVIRONMENT

1.1 Definition of the radio network system

The objective of this handbook is to present an efficient radio planning process for cellular radio networks and to discuss different solutions for detailed planning activities in order to achieve the most cost-efficient high quality radio network. The planning activities presented are based on the Pan-European GSM (Global System for Mobile Telecommunication) network requirements but most of them can also be adapted for the radio planning of WCDMA (Wideband Code Division Multiple Access) that was selected as the air interface technology for the third generation UMTS (Universal Mobile Telecommunication System) network architecture. The basic radio planning is very similar for different radio systems, thus the digital GSM radio network is used as a reference system in this book and the detailed radio planning differences for the GPRS and UMTS are explained separately.

The GSM mobile network has three main subsystems, which are called a *network subsystem* (NSS), *network management system* (NMS) and a *base station subsystem* (BSS) all of which are illustrated in Figure 1.1.

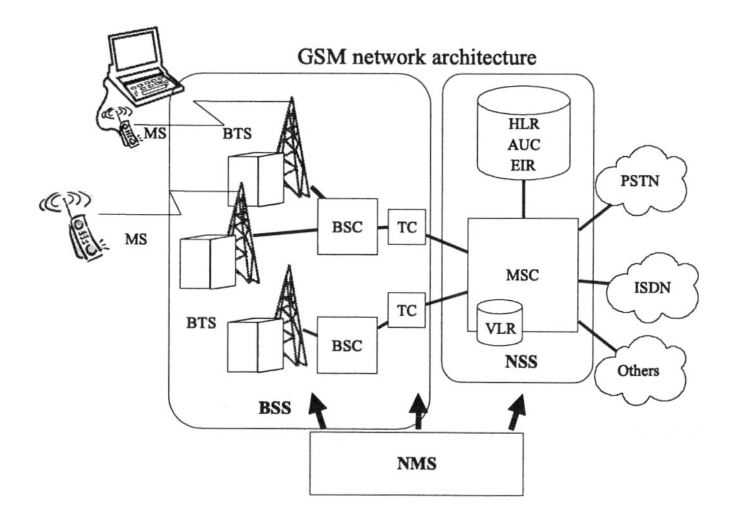

Figure 1.1. The GSM mobile network subsystems.

The network subsystem contains a switching part of the GSM system as a mobile switching centre (MSC) and gateway switches (G-MSC) that are the connection points between the GSM mobile networks and for example PSTN (public switch transmission network) networks. The network management system contains equipment for the mobile network operation and maintenance as an *operation and maintenance centre* (OMC). Finally, the base station subsystem contains the equipment to manage the radio interface between the mobile station and cellular radio network. The *base station controller* (BSC) found between *a base transceiver station* (BTS) (also called a *base station,* BS) and a MSC controls the main functions of the mobiles (called also *mobile station*, MS) and base stations. The base stations have a radio connection with the mobile stations and they represent a physical radio interface of the mobile network as illustrated in Figure 1.1. Thus, the base station has to be capable of communicating with the mobile stations over a certain coverage area and to offer enough capacity (traffic channels) for this communication (including the signalling and data transfer that can be speech or data transmission). It has to be noted that the base station as a term means originally *the base station equipment* itself that may typically manage 1–6 independent coverage areas which each contain an independent group of *transceivers* (otherwise *frequencies*). These coverage areas can be provided by using omnidirectional or directional antennas for transmission and reception and they can also be called *cells*. Figure 1.2 introduces *a base station site* with one cell (omnidirectional transmission and receiver antenna), with two cells called *2-sector* (wide beam directional antennas), and with three cells called *3-sector* (narrow beam directional antennas).

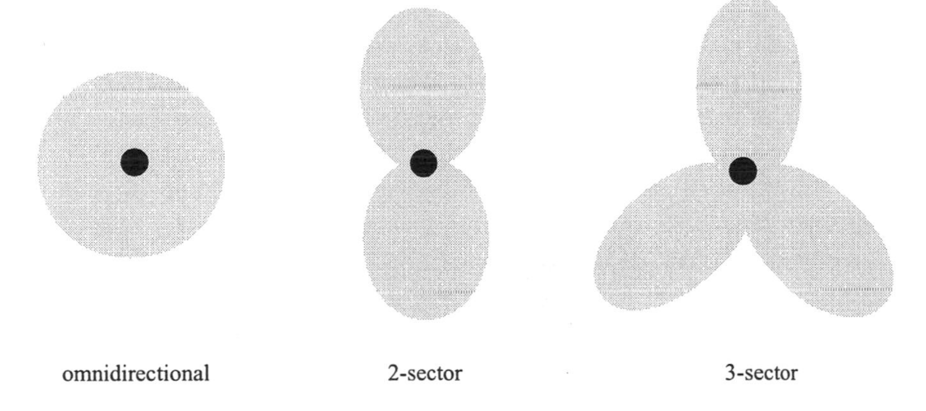

Figure 1.2. Base station site configurations of 1–3 cells.

Base station as a term is inadequate to describe all related issues and therefore this books adopts the following terminology and terms in order to speak accurately of the correct items:

Base station site	= the physical location of the base station and the total configuration of the base station equipment and antenna lines at that location
Base station equipment	= the equipment without antenna line (hardware plus software)
Cell or sector	= the physical coverage area of one base station antenna direction.

Thus the configurations in Figure 1.2 all represent a base station site that only has different configurations of the base station equipment and antenna line.

Omnidirectional cells are usually not implemented because their coverage and capacity properties are the worst. The 2-sector base station is typically used for road coverage and 3-sector configuration for both urban and rural areas as it provides the largest coverage areas (because of the high gain property of the directional antennas) and also the highest capacity (from the directional antennas).

1.1.1 Coverage terminology and definitions

The main function of a base station is to transmit and receive voice or data messages to and from mobile stations (MS) by using the radio interface. In order to have a continuous connection between the BTS and MS the maximum distance between them have to be defined. This definition is based on the minimum possible *signal-to-noise-ratio* (SNR) at both receiving ends and also on the maximum transmission power at both transmission ends. Both receiving and transmission levels can be improved by using different technologies such as low noise amplifiers (reception), diversity reception (reception, transmission), high gain directional antennas (reception, transmission) and power amplifiers (transmission).

The difference of the maximum transmission and minimum receiving levels between the BTS and MS antenna elements is called the *maximum path loss*. It defines the maximum allowed attenuation between the BTS and MS and vice versa. The maximum distance between BTS and MS can then be defined when the *radio propagation* (attenuation versus distance, typical unit is dB/dec) is known in a certain *propagation environment.* The attenuation is higher in a built-up environment compared to a rural environment (with no man-made obstructing constructions). The slope of the radio wave attenuation as a function of distance is called a *radio propagation slope* and that has a strong effect on the maximum distance between the BTS and MS. The propagation slope depends heavily on the propagation environment and also on antenna height which is thus a critical parameter for coverage planning.

1.1.2 Capacity terminology and definitions

The number of connections (otherwise calls) between the base station and mobile stations is an essential parameter when the mobile network configuration is designed. *The main target in radio network infrastructure planning is to minimise the number of base station sites in order to minimise their construction, operational and maintenance costs.* The number of base stations is minimised by maximising the number of connections per base station which brings enormous cost savings. The number of connections between the BTS and MS depends on the number of *frequencies* and how often these frequencies can be reused in a certain area and between base stations. The maximum number of frequencies in the GSM900 (GSM at 900 MHz band) system is 125 in the *uplink* (MS→BTS) and *downlink* (BTS→MS) directions. Each frequency is also called *a channel* and thus there are a maximum of 125 channels available for both directions. Each of these channels can be reused at some distance from the base station. This distance describes the number of neighbouring base stations where some other frequency has to be assigned. This group of base stations forms a cluster structure as depicted in Figure 1.3.

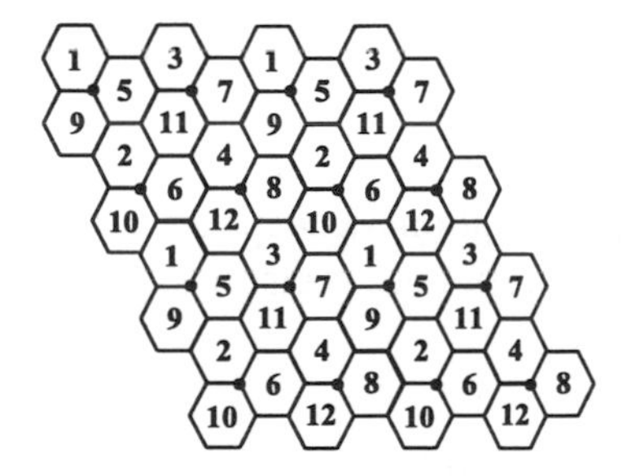

Figure 1.3. An example of the frequency (planning) cluster structure in a cellular network.

Twelve different frequencies or frequency groups in Figure 1.3 have to be used in order to avoid interference. This value of twelve is also called a *frequency reuse factor* or *number* because it defines the period of the frequency reuse. Minimum frequency reuse factor can be calculated by analysing the *co-channel carrier-to-interference ratio* (*C/I*)—the higher the minimum *C/I* the lower the frequency reuse factor. Moreover, the minimum *C/I* value depends mostly on the radio propagation environment. If the maximum configuration was applied and all 125 frequencies were available, 125/12 = 10.4 frequencies on average at each cell could be used. This is also the maximum configuration or maximum capacity of each cell and the total capacity of the whole network depends on the number of cells. The maximum average number of BTS–MS connections at each cell can then be calculated by multiplying 10.4*8 = 83.2. The multiplication by 8 is based on the GSM being a time division multiple access (TDMA) based system where 8 BTS–MS connections (otherwise *time slots*) can be provided by using the same frequency because these connections alternate, as can be seen in Figure 1.4.[1] Each of these time slots is called a *traffic channel* because a full data or voice call can be performed on any of these channels.

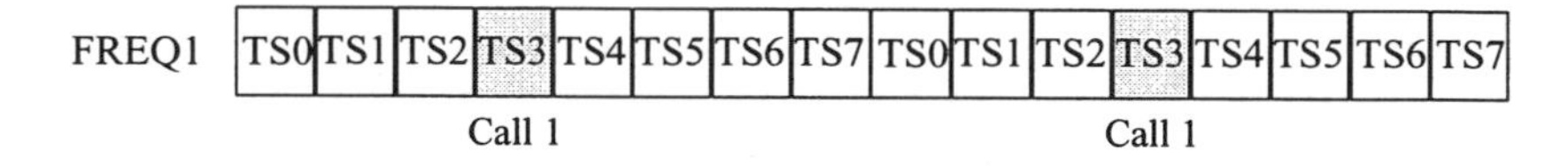

Figure 1.4. The GSM channel structure.

It has been noted that the antenna height defines the base station coverage area and also the capacity requirements for each base station via the number of base stations required to cover the planning area. If the coverage and capacity requirements can be combined and the optimised values for the base station antenna heights can be defined in the radio planning over several years, a cost-efficient radio network infrastructure can definitely be achieved.

1.2 Cellular radio network

1.2.1 Cellular concept

The radio interface in mobile cellular networks is based on a cellular approach because frequencies are required to be reused due to a limited amount of frequency resources and the large number of users sharing these resources. The cellular infrastructure in a suburban area is depicted in Figure 1.5. Each cell contains the base station equipment, which transmits and receives voice and/or data traffic by using an antenna array and a certain group of frequencies. All frequencies are reused after a sufficient required distance to avoid too high interference levels (Figure 1.5). The minimum distance between the same frequencies is determined by the environment—hills, trees, buildings, etc.—effecting radio propagation.

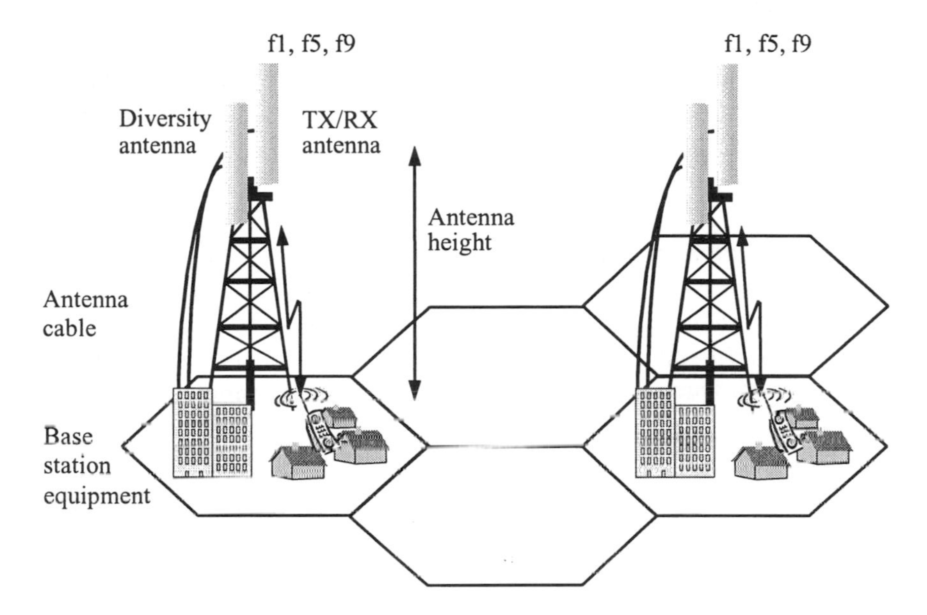

Figure 1.5. Base station equipment and especially the base station antenna array type and height determine capacity in a cellular network.

1.2.2 Capacity, coverage and quality of service

In addition to the propagation environment radio propagation also strongly depends on the type of base station antenna array used and on its effective height with respect to the ground and surrounding environment. If the base station antenna array position is too high (e.g. clearly above the average building height in an urban area), radio waves propagate too far causing interference and the cellular network will be easily *capacity limited* because these frequencies can not be reused in the same area due to the interference. Thus, the radio propagation environment and the base station antenna array configuration determine the maximum frequency reuse over a network coverage area. Furthermore, the maximum frequency reuse specifies the maximum capacity and the minimum number of required base stations over a cellular network service area from a capacity point of view.

However, capacity requirements alone do not determine the Quality of Service (QOS) in the mobile networks. Quality of Service also involves location probability at a base station coverage area. Location probability determines the likelihood of exceeding a certain threshold power or field strength level over a base station coverage area in indoor or outdoor locations depending on the operator's requirements. Location probability depends again on the base station antenna array height—the higher the antenna array position the higher the location probability at the same base station coverage area.

1.2.3 Optimal solution

Having introduced the links between base station antenna height and coverage and capacity, it has been noted that the link between the antenna height and radio propagation is a key element when coverage and capacity are designed for cellular network infrastructures. Radio propagation is environment-dependent and the performance of different coverage and capacity solutions are thus related to factors like *morphography type* and *topography* or building height. The influence of these environmental aspects has to be emphasised when radio planning applications are considered. The optimal solution for coverage is not necessarily the optimal solution for capacity and interference, and vice versa, but the optimal solution is something where coverage, capacity and interference solutions are combined.

Cellular radio planning is always an optimisation task where both coverage and capacity have to be maximized and interference has to be minimised. As in all optimisation tasks the biggest question is where to start the process. Traditionally it is defined that radio planning has to be started from the coverage predictions in order to estimate the number of base stations over a certain coverage area. Next follows an evaluation of additional base stations needed for capacity. However, this approach fails to prioritise the antenna height definition process in Figure 1.6. Antenna height is to be prioritised more than the other parameters, as it has a significant effect on coverage predictions and it is also a major parameter when capacity and interference are optimised. Thus, the antenna height definition is an optimised starting point for radio network system planning, see Figure 1.6.

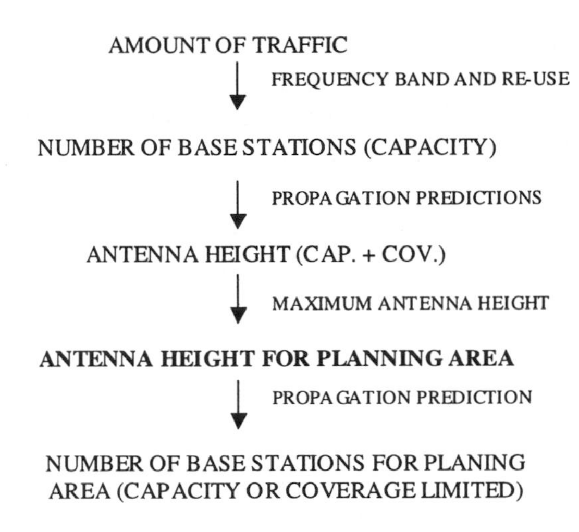

Figure 1.6. The antenna height definition process.

As is shown in Figure 1.6, the antenna height defines the base station coverage area—the higher the antenna position the larger the base station coverage area—and it also defines the capacity—the lower the antenna position the more often frequencies can be reused—thus optimised antenna height gives both required coverage and capacity.

1.3 Radio propagation environment

Radio propagation (and thus coverage and capacity) depends on the propagation environment which is typically divided into three major classes—*urban, suburban* and *rural*—and into two special classes—*microcellular* and *indoor*. Area types—urban, suburban and rural—refer to constructed (e.g. buildings) or natural (e.g. trees) obstacles, which vary in size and density around the base station and mobile station antennas. In the major classes the environment is called *macrocellular* when the *base station antenna arrays are above the average rooftop level* and the radio propagation environment is called *microcellular* when the base station antenna array is implemented below the average rooftop level (see Figure 1.7).

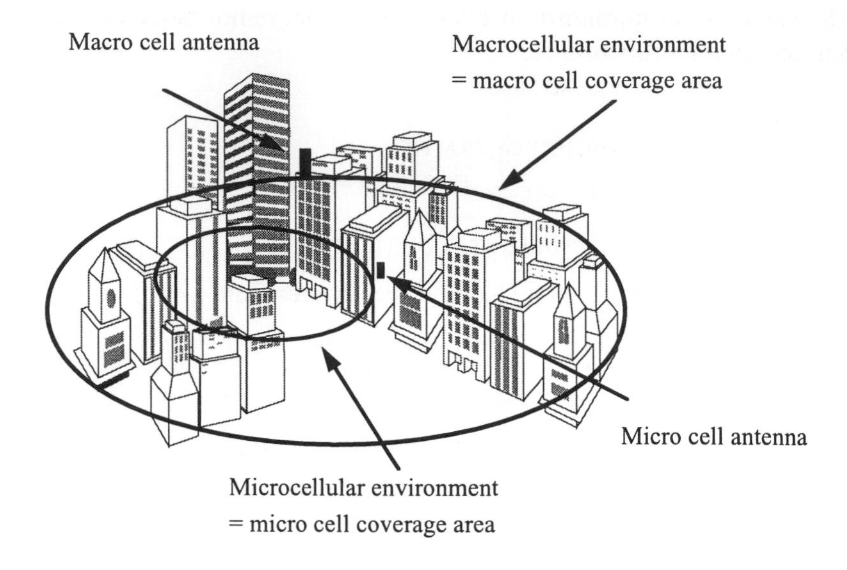

Figure 1.7. Macrocellular and microcellular radio propagation environment.

Thus, microcellular environments only exist in urban areas where there are buildings higher than three or four floors. Micro base stations (a base station in a microcellular environment) are specifically required in high capacity areas where frequencies have to be reused several times. When antenna arrays are below the rooftops, surrounding buildings prevent propagation, base station coverage area is small and frequencies can be reused more often. Moreover, macrocellular, microcellular and indoor environments each have individual radio propagation channel characteristics. Thus, the radio wave propagation in each environment has to be studied separately.

1.3.1 Radio propagation parameters related to the environment

It can be concluded that the radio propagation environment is specified by the base station and mobile station antenna environments based on the:

- Morphography type (urban, suburban, rural)
- Antenna location above or below the average rooftop level (macrocellular, microcellular)
- Mobile station location (outdoor or indoor).

In each of these environments the radio wave propagation can be described by the following parameters:

- Angular spread
- Delay spread
- Fast and slow fading properties
- Propagation slope.

These parameters define the *characteristics of radio wave propagation* in the different environments and also for different systems (for example GSM and UMTS). Moreover, these parameters have a significant effect on the base station coverage and capacity planning in all environments and they define the optimised coverage and capacity solutions in all different environments. Each propagation environment can thus be categorised based on the radio propagation parameters listed above and each of these parameters have nominal values in each propagation environment.

1.3.1.1 Angular spread

Angular spread describes the deviation of the signal incident angle. Usually we are interested in the signal coming to the base station antenna. Angular spread can be calculated based on the incident angle of the received power in the horizontal and vertical planes:[2]

$$S_{\Phi} = \sqrt{\int_{\overline{\Phi}-180}^{\overline{\Phi}+180} (\Phi - \overline{\Phi})^2 \frac{P(\Phi)}{P_{\Phi\_total}} \mathrm{d}\,\Phi} \qquad \textit{Equation 1.1}$$

where $\overline{\Phi}$ is the mean angle, $P(\Phi)$ is the angular power distribution and P_{Φ_total} is the total power.

It has to be noted that the angular spread can be calculated in two planes—either horizontal or vertical. The received power from the horizontal plane is still the most important because of obstructing constructions: most of the reflecting surfaces are related to the horizontal propagation and thus multiple BTS–MS propagation paths exist more in the horizontal plane. The horizontal angular spread is around 5–10° in macro cells and very wide in microcellular and indoor environments because the reflecting surfaces surround the base station antenna. The angular spread has a significant effect on antenna installation direction and on the selection and implementation of traditional space diversity reception. Vertical angular spread additionally influences, for example the base station antenna array tilting angle when the co-channel interference is reduced. The angular spread is also a key parameter when the performance of the adaptive antennas is discussed because the optimisation of the carrier-to-interference ratio (*C/I*) depends strongly on the incident angles of the carrier and the interference signals. Thus, the performance of the adaptive antennas is lower or more difficult to achieve in the microcellular environments than in the macrocellular environments.

1.3.1.2 Multipath propagation and delay spread

The signal propagates between BTS and MS via multiple different paths due to reflections, diffractions and scattering. This is called *multipath propagation* (Figure 1.8).

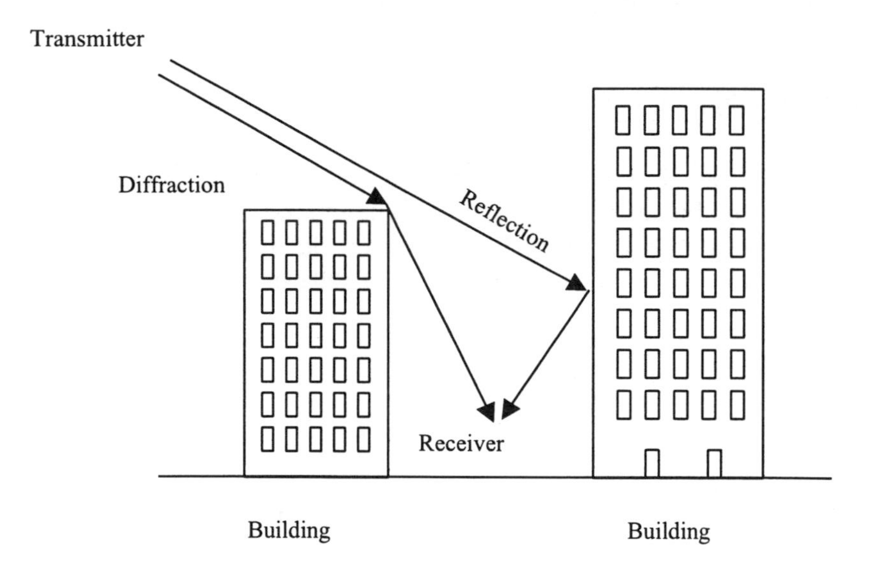

Figure 1.8. Multipath propagation.

These multiple paths have different propagation lengths and thus the accurate reception time of the different multipath signals varies. The amount of this time variation can be measured as *delay spread.* The delay spread S_τ can be calculated from the power-delay profile of the radio channel $P_\tau(\tau)$, which is the received power as a function of delay.[3]

$$S_\tau = \sqrt{\frac{\int_0^\infty \left(\tau - \overline{\tau}\right)^2 P_\tau(\tau)d\tau}{P_{\tau\_tot}}}$$ *Equation 1.2*

where $\overline{\tau}$ is the average delay and P_{τ_tot} is the total received power.

Multipath propagation is typically not harmful for GSM receiver performance because the GSM standard requires receivers to be equipped with an equaliser which can equalise for path delay differences up to 16 µs. This corresponds to the 4.5 km path length difference. If the path delay difference is larger than 16 µs (typically reflections from hills in a rural area) the delayed signals cause co-channel interference *I*. The amount of co-channel interference depends on the strength of the long reflections.

The equaliser of the GSM receiver also makes it possible to build-up so-called distributed antenna systems (DAS). This is especially useful for indoor locations where base station coverage area is provided by *N* base station antenna arrays that all transmit and receive the same data as illustrated in Figure 1.9.

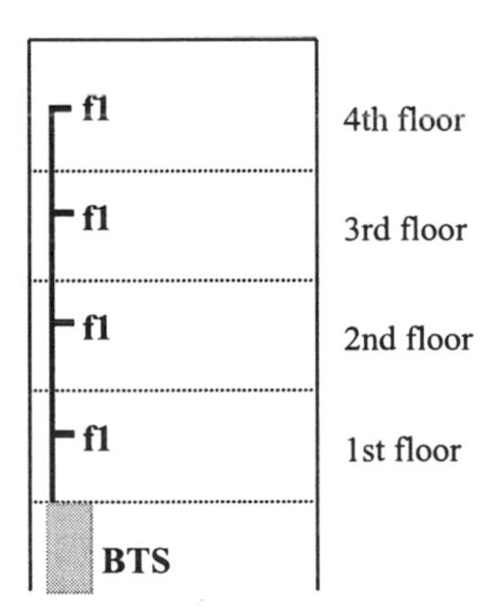

Figure 1.9. Distributed antenna system (DAS).

The values of delay spread are related to the environment and the largest values in a macrocellular environment can be found in hilly rural areas where the reflecting surfaces can be very far away. In the macrocellular urban environment typical values are less than in hilly rural areas. The smallest values of delay spread have been measured in special environments like microcellular and indoor. These different delay spread values in the different environments do not cause any problems in basic GSM coverage or capacity planning. However in detailed coverage and capacity planning, when different radio planning concepts, methods, and technologies are required to achieve coverage and capacity in excess, the multipath propagation has to be addressed more carefully.

The delay spread has a strong effect for example on the performance of frequency hopping because the required separation of the frequencies to provide uncorrelated signals—called coherence bandwidth Δf_c—is a function of the delay spread.[4]

$$\Delta f_c = \frac{1}{2\pi S_\tau}$$ *Equation 1.3*

The coherence bandwidth depends on the environment and it should be noticed that a very wide and impractical frequency separation is required in microcellular and indoor locations and thus the applicability (and the performance) of the frequency hopping is limited in these environments.

1.3.1.3 Fast fading

The mobile station or base station receives in one moment the same signal arriving via different radio paths as mentioned in the previous section. The total received signal is a contribution of all the arrived signal multipath components based on the superposition principle. The different signal components, arriving via different radio paths, have different amplitude and phase due to the different lengths of the radio path and different reflection and diffraction properties. Thus, the sum of the received signals can be constructive or a destructive depending on the phases of the multipath components.

The amplitude and especially the phase of the multipath components change very quickly when the mobile station moves. A movement Δd over distance of one signal wavelength (30 cm at 900 MHz) causes a phase change $\Delta\phi$ of up to 2π depending on the direction of the movement in respect to the direction of the incoming multipath component α (Equation 1.4).

$$\Delta\phi = -\frac{2\pi}{\lambda}\Delta l = -\frac{2\pi\Delta d}{\lambda}\cos\alpha \qquad \textit{Equation 1.4}$$

Because of these fast changes in the phases of the multipath components, the phase and amplitude of the total signal also changes rapidly when the receiver moves. This is called *fast fading*. When all of the components have random uniformly distributed phase (0, 2π), the amplitude of the total signal is Rayleigh distributed [4] (fast fading is also called Rayleigh fading). This occurs when there is no single dominating signal path, for example no-line-of-sight component (NLOS). In a Rayleigh fading situation the signal amplitude variations can be very large, see Figure 1.10.

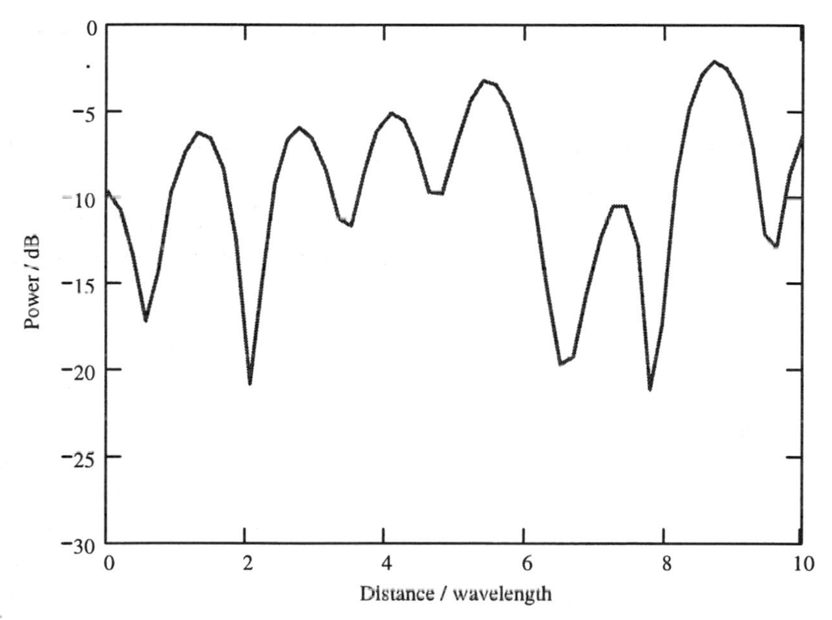

Figure 1.10. Power of Rayleigh faded signal as a function of distance.

The Rayleigh distribution can be described by the cumulative density function (CDF). This is shown in Equation 1.5.

$$P(r \le R) = \int_0^R \frac{r}{\overline{r^2}/2}\exp\left(-\frac{r^2}{\overline{r^2}}\right)\mathrm{d}\,r \qquad \textit{Equation 1.5}$$

where r is the envelope of the fading signal and $\overline{r^2}$ is the mean square value of the fading signal.

When a signal in one incoming direction has a higher amplitude than the others, as in the case of line-of-sight (LOS), the total signal amplitude is no longer Rayleigh but Rician distributed and the fading is called Rician fading. Rician fading happens specifically in micro base stations where the mobile station is close to the base station antenna array and one signal path (the LOS component) is clearly better than the others. The cumulative density function of Rician distribution is shown in Equation 1.6.

$$P(r \le R) = \int_0^{R_0} r_0 \exp(-\frac{r_0^2 + a_0^2}{2}) I_0(a_0 r_0)\, \mathrm{d}\, r_0 \qquad \textit{Equation 1.6}$$

where

$$r_0 = \frac{r}{\sqrt{\overline{r^2} / 2}} \qquad \textit{Equation 1.7}$$

$$a_0 = \frac{a}{\sqrt{\overline{r^2} / 2}} \qquad \textit{Equation 1.8}$$

$$R_0 = \frac{R}{\sqrt{\overline{r^2} / 2}} \qquad \textit{Equation 1.9}$$

The variable a is the amplitude of the direct (dominating) signal. A special case of Rician distribution is Rayleigh, in which case the direct signal amplitude is zero ($a = 0$).

Fast fading distribution varies depending on the BTS–MS connection (LOS or NLOS situations). In addition, the environment has an effect on the received power distribution especially in microcellular and rural situations. In a microcellular propagation channel, as well as in rural situations, there may be several major signal components and thus field strength distribution is not constant for different situations.

The amplitude of the signal at the receiver via two different radio channels is shown in Figure 1.11. When these signals are combined together at the receiver an additional 4–5 dB improvement can be achieved. This multiple reception is called diversity or diversity reception and it is based on the availability of uncorrelated signals at the receiving end. These uncorrelated signals (as in Figure 1.11) can mainly be provided by the base station antenna configuration (antenna diversity), or by frequency or time separation of the signals.

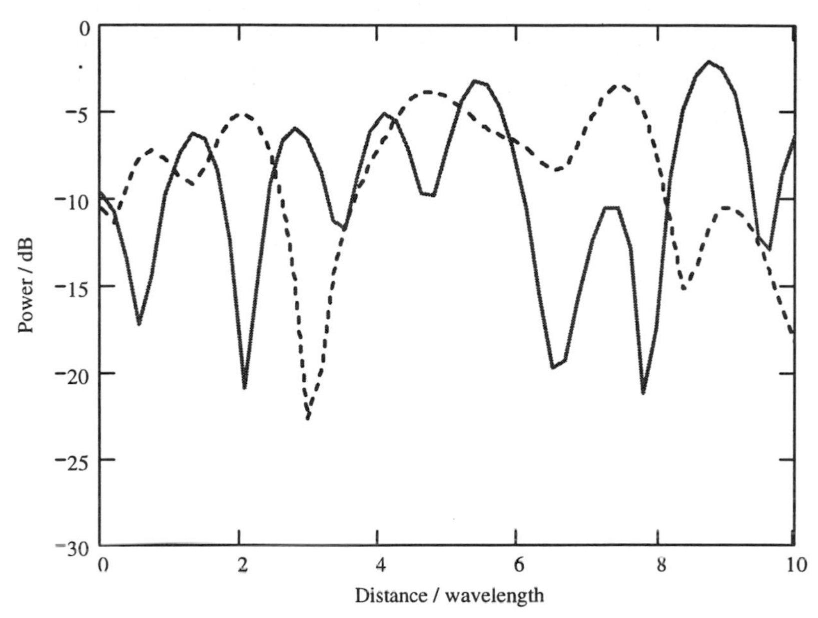

Figure 1.11. Amplitude of a signal received via two different multipath radio channels.

1.3.1.4 Slow fading

Slow fading or log-normal fading is the variation of the local mean signal level over a wider area, and has been observed by Young.[4] The local mean is the mean value of the Rayleigh or Rician fading signal amplitude. This log-normal fading is caused by the obstacles (buildings, trees, etc.) that change the average received signal level and thus bring about shadowing. The variation of the signal amplitude local mean value over the wider area is log-normally distributed and thus it is called log-normal fading.

Normally in cellular radio planning, cell range and area are predicted with a target location probability factor (for example 95 percent area probability) indicating the statistical properties of the signal strength over the cell area. In order to achieve the required location probability level some margin has to be taken into account in the planning. Moreover, these margins are usually applied for macro base station coverage planning but the slow fading margin requirements for micro base station coverage planning should be considered separately; the majority of BTS–MS connections in micro base stations are in LOS situations or NLOS only from around the corner (as seen in Figure 1.7). Thus, the mobile stations are so close to the base station that no typical shadowing arising from obstacles occurs and therefore the shadowing margin for the micro base stations' coverage is basically not needed.

1.3.1.5 Propagation slope

Radio wave attenuation in free space is proportional to the square of the distance r (in dB scale 20 dB/dec).[5] In communication links between the base station and mobile station in a macrocellular environment the signal levels typically decrease 25–50 dB/decade (that is, the propagation slope is 25–50 dB/dec) depending on the terrain type (morphography and topography), as in Table 1.1.[5]

Table 1.1. Typical propagation slopes for the different environments.

Environment type	Slope (dB/dec)
Microcellular	20
Rural	25
Suburban	30
Urban	40
Dense urban	> 45

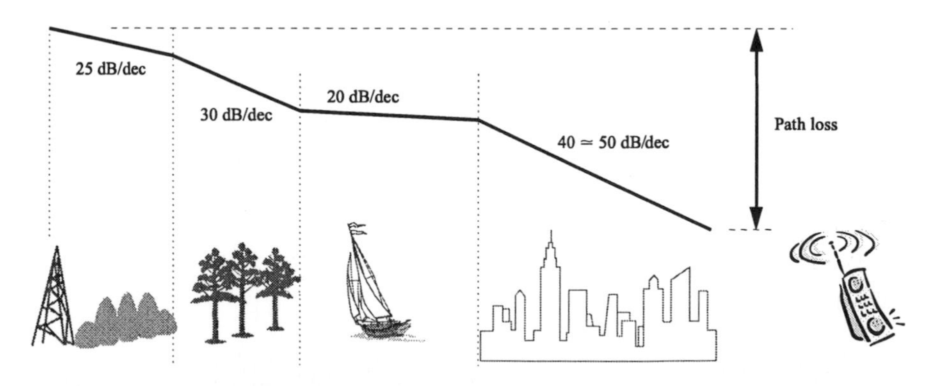

Figure 1.12. The propagation slope and path loss.

The propagation slope has a strong effect on the path loss which is the attenuation between the transmission and receiving ends, see Figure 1.12.

The path loss can also be determined with path loss exponent by using the equation

$$L = L_0 d^{\exp}$$ *Equation 1.10*

where L is path loss, L_0 is the path loss at the reference point, d is the distance between the BTS and MS and *exp* is the path loss exponent which equals the propagation slope divided by 10 (*exp* = *slope*/10). Notice however that the propagation slope is not constant over the BTS–MS link even if the propagation environment is constant. Close to the base station the propagation slope follows the inverse square law according to the two-ray equation, excluding fading dips. Thus, the propagation slope near the transmitting antenna is lower than the propagation slope at greater distances. The distance of the propagation slope change is called the "breakpoint distance" and it can be calculated by using the equation [5]

$$B = 4\frac{h_{BTS} h_{MS}}{\lambda}$$ *Equation 1.11*

where h_{BTS} is the height of antenna 1 (BTS) and h_{MS} is the height of the other antenna (MS).

The existence of the propagation slope breakpoint is an important element that should be taken into account in cellular radio network planning. The idea is to plan the radio network in such a way that a mobile station (MS) connected to a certain BTS experiences a lower propagation slope to the serving BTS and a greater propagation slope to any interfering BTS. Hence, this leads to good coverage and a low level of interference at the same time. Of course, a realisation of this idea in an actual cellular radio network is not very easy in practice. In Figure 1.13 the breakpoint distances at different frequencies are given as a function of the BTS antenna height when the MS antenna height is 1.5 m.

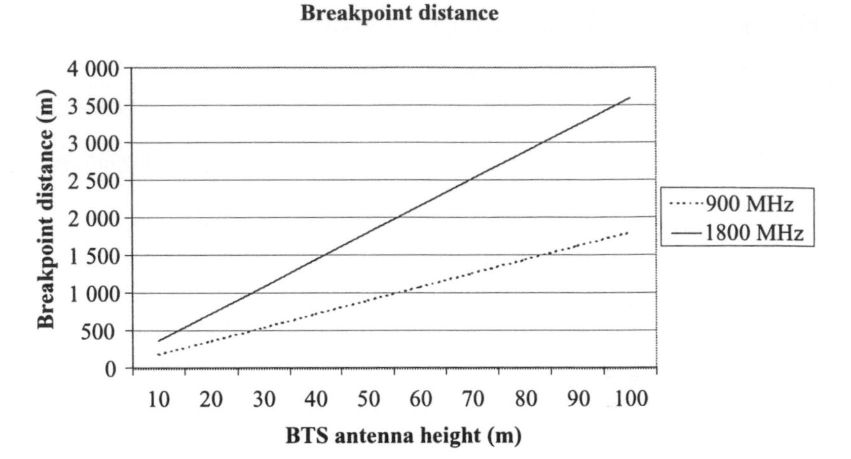

Figure 1.13. The breakpoint distances of the frequencies 900 and 1800 MHz as a function of the BTS antenna height (h_{MS} = 1.5 m).

It can be observed from Figure 1.13 that the breakpoint distances at 900 MHz are quite short and therefore it is difficult to benefit the phenomenon of a lower propagation slope at closer distances at this frequency. Whereas at the 1800 MHz frequency band the breakpoint distance can be increased quite rapidly by increasing the BTS antenna height. This makes it possible to select the BTS antenna height in order that lower path loss in desired areas may actually be reached.

The propagation slope and breakpoint distance have to be engaged in a macrocellular propagation environment and it is especially important in urban areas where coverage should be maximized together with capacity.

1.3.2 Characteristics of radio propagation environments

The typical values for the different environments of the aforementioned radio channel characteristics are gathered in Table 1.2 at 900 MHz.

The figures in Table 1.2 show the critical issues that have to be taken into account in radio system planning. In the microcellular and indoor environments there are many LOS connections and there are often several predominant paths between BTS and MS and thus the fast fading statistics vary.

Table 1.2. Characteristics for different radio propagation environments at 900 MHz.[2–7]

	Angular spread (°)	Delay spread (μs)	Fast fading	Slow fading standard deviation (dB)	Propagation slope (dB/dec)
Macro-cellular					
Urban	5–10	0.5	NLOS	7–8	40
Suburban	5–10		NLOS	7–8	30
Rural	5	0.1	(N)LOS	7–8	25
Hilly rural		3	(N)LOS	7–8	25
Micro-cellular	40–90	<0.01	(N)LOS	6–10	20
Indoor	90–360	<0.1	(N)LOS	3–6	20

Second, the propagation slope shows that the attenuation varies strongly in different environments. This variation is still more for the macrocellular area type comparison (urban/suburban/rural) because the propagation slope is not required so much in microcellular and indoor planning. In a microcellular environment the propagation slope is typically around 20 dB/dec and street "canyons" are of special concern. Tunnels are a special case because the propagation slope depends on the tunnel shape (rectangular or arched) and on the polarisation (vertical or horizontal).[8] Hence, radio coverage planning and the propagation characteristics have to be considered carefully in tunnel applications.

Finally the influence of the angular spread and delay spread on the different propagation environments has to be understood. These parameters have the strongest effect on radio system planning because they define the application of frequency hopping and adaptive antenna concepts and the antenna configurations of the diversity reception. Figures of the angular spread show that the microcellular and indoor areas are totally different environments compared to the macrocellular. It can be concluded that the microcellular and indoor environments are the most difficult ones to improve the system efficiency or to reduce the interference level, for example by using adaptive antennas. Moreover, the advantage of the wide angular spread in a microcellular environment can be seen in the space diversity reception which can be implemented by using a very compact antenna configuration.

The same microcellular and indoor locations cause problems because the delay spread is so short. The short delay spread means that frequency separation in the frequency hopping has to be wide and therefore this method is impractical to use to combat fast fading in microcellular and indoor locations. Hence, diversity reception is needed to improve and balance the reception level in these environments.

1.4 Network evolution path

A radio network has to be extended throughout its evolution because of the increase in traffic (capacity extensions) and demands of better indoor and outdoor coverage (coverage extensions). Radio network evolution can be divided into different phases, using four main categories:

1. Basic coverage deployment
2. The first capacity extensions and coverage enhancement deployment (indoor coverage)
3. Intelligent capacity enhancement deployment with strong indoor coverage improvement
4. Base station coverage area splitting/microcellular/dual band deployment.

These different *radio network evolution phases* are also related to the different *radio propagation environments*. In the early phases base stations are more of a macrocellular type but after base station splitting the base station antennas should be installed in lower and lower positions in order to avoid too large a coverage areas, simultaneously the propagation environment also changes towards microcellular. The use of low antenna positions improves the capacity but simultaneously the coverage is significantly reduced and more and more base stations are needed to fill the new coverage holes. Thus, the key point in radio network evolution is to keep the antennas high by utilising initially all possible coverage and capacity related functions. Simultaneously, the coverage and capacity limitations of each radio propagation environment have to be understood (based on the radio environment characteristics in Table 1.2) in order to be able to optimise the base station coverage and capacity at each environment.

Radio network evolution phases are thus always related to a specific radio network configuration that includes;

- A specified average base station antenna height (macro or micro base stations)
- Required capacity and quality related hardware and software features.

The different evolution phases always occur in succession which is called a *radio network evolution path* (Figure 1.14).

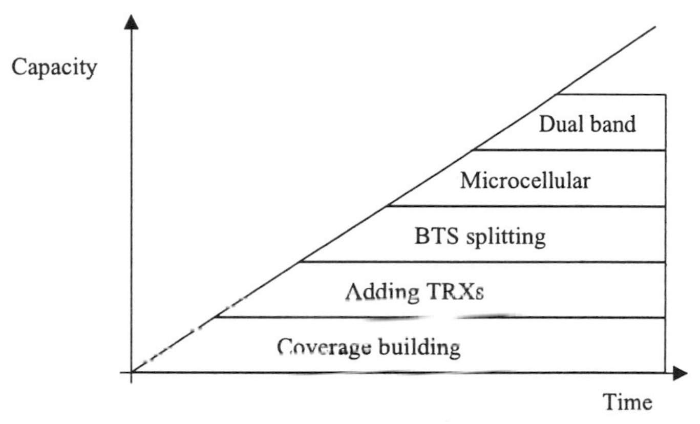

Figure 1.14. The radio network evolution path.

Figure 1.14 shows that in the beginning the coverage has to be good enough which defines the minimum number of base stations for a certain area—the radio network is coverage limited. After the coverage requirements have been exceeded by the capacity requirements, it starts to influence the radio network configuration—the radio network is capacity limited.

First the different capacity related hardware and software features are utilised and finally more cells are implemented (first by splitting the macro base stations and finally the micro base stations). The utilisation of the new frequency band—implementation of the dual band network—is one of the last items to be covered in the capacity enhancement process as it is difficult to acquire new frequencies and also because this causes some difficulties in planning. Figure 1.14 shows when different propagation channel characteristics (related to the different environments and different evolution phases) are to be concerned. When there are enough base stations in the radio network the capacity related functions can be utilised again and thus *the radio network evolution process continues in a new environment with new radio propagation characteristics in a similar way as at the beginning*.

1.5 Conclusions

Having introduced the GSM system and terminology for radio interface, the philosophy of the cellular radio network and the key links between the base station antenna height, radio propagation and radio propagation environments were defined. These definitions formed the basis for discussion about radio propagation environments and their different characteristics and influence on radio system planning. Table 1.3 reviews the conclusion that can be made: that traffic demand from the radio network and the average base station antenna height are the parameters that have to be optimised in order to maximise the overall radio quality called Quality of Service (the coverage, capacity, interference level and cost efficiency). Hence, an optimised radio network configuration can be provided by taking into account the characteristics of radio wave propagation in different environments.

Table 1.3 The highlights of the radio propagation environment.

Subject	Findings
Radio "system" planning	–Coverage and capacity have to be planned together –Antenna height links the coverage and capacity
Base station antenna height	–The antenna height has the strongest and direct effect on the coverage, capacity and interference
Radio propagation environment	–Macrocellular/Microcellular –Urban/Suburban/Rural –Outdoor/Indoor
Angular spread	–Narrow in macrocellular environment –Wide in microcellular environment –Influence on the antenna configuration
Delay spread	–Long in macrocellular environment –Short in microcellular environment –Influence on the frequency hopping
Fast fading	–Effect can be reduced by using diversity reception
Slow fading	–Because of the obstacles as buildings
Propagation slope	–Defines the environment type or vice versa
Breakpoint distance	–Base station coverage areas should be planned based on the breakpoint distance

1.6 References

[1] ETSI, Digital cellular telecommunications system (Phase 2+), Mobile radio interface layer 3 specification, GSM 04.08.

[2] Jaana Laiho-Steffens, "Two dimensional characterisation of the mobile propagation environment," Licentiate's Thesis, Helsinki University of Technology, 1996.

[3] D. Parsons, "The Mobile Radio Propagation Channel," Pentech Press, 1992.

[4] W.C. Jakes, Jr., (ed.), "Microwave Mobile Communications," Wiley-Interscience, 1974.

[5] W.C.Y. Lee, "Mobile Communications Design Fundamentals," John Wiley & Sons, 1993.

[6] W.C.Y. Lee, "Mobile Cellular Telecommunication Systems," McGraw-Hill Book Company, 1990.

[7] S. Saunders, "Antennas and Propagation for Wireless Communication Systems," John Wiley & Sons, 1999.

[8] Jyri S. Lamminmäki, Jukka J.A. Lempiäinen, "Radio Propagation Characteristics in Curved Tunnels," IEEE—Microwaves, Antennas and Propagation, August 1998, vol. 145, no. 4, pp. 327–331.

[9] ETSI, Digital cellular telecommunications system (Phase 2+), Radio subsystem link control, GSM 05.08.

Chapter 2

RADIO SYSTEM PLANNING PROCESS

2. RADIO SYSTEM PLANNING PROCESS

2.1 Radio system planning phases

The main parameters—antenna type, antenna height, area type—that effect *radio system planning* also define the radio propagation environment that specifies the characteristics of the radio propagation which furthermore has a significant effect on the coverage and capacity in the radio network. Radio system planning is a process that defines the stages—visits in the area, measurements, planning, documentation—required to provide a desired *radio network plan* for a certain geographical area. Moreover, the radio system planning process has to be defined carefully and carried out in different phases in order to manage the strong influences between

- *coverage*
- *capacity*
- *quality* (interference probability).

These three areas must all be optimised in order to achieve a cost-efficient and overall high Quality of Service (containing inter alia good speech quality, minimum radio network congestion, minimum number of drop calls or handover failures) radio network. In coverage planning the aim is *to maximise the base station coverage areas* and thus minimise the required infrastructure. Correspondingly, the base station need has to be minimised in the capacity planning *by reusing the frequencies as often as possible*. These two topics have different applications: base station coverage area can be maximised by *maximising the base station antenna height* and the base station capacity can be maximised by maximising the frequency reuse that can be reached by *minimising the base station antenna height*. Furthermore, quality is not a real planning topic but is a very important "issue" and refers to primarily interference that can be connected to capacity and frequency planning that depend on coverage issues like the base station antenna heights. *Quality connects radio network coverage and capacity planning and is related to frequency planning*. This illustrates that any of these "topics" can not be maximised but that they all have to be optimised (otherwise, system planning) in order to achieve a cost-efficient and high Quality of Service radio network.

In order to plan good coverage simultaneously optimising capacity and maximising quality, the radio system planning process and key parameters for this process have to be clearly defined. Figure 2.1 presents the radio system planning process and its different phases which can be adapted to radio networks from commencement of deployment to their radio evolutionary extension. The same phases and studies are required time after time to deploy new and maintain existing networks. Three major radio system planning phases:

- *dimensioning*
- *detailed radio system planning*
- *optimisation*

can be identified and each of these has a specific purpose. First, *dimensioning* is required to "generally" analyse the network configuration and to decide the radio network deployment strategy. Next, the radio network is accurately designed in the *detailed radio system planning* phase and finally the radio network evolution requirements are considered in the *optimisation and monitoring* phase.

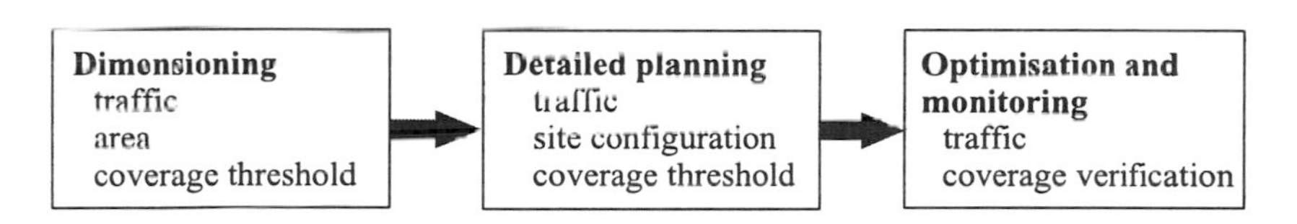

Figure 2.1. Radio system planning process phases and their key parameters.

Figure 2.1 shows that the three major radio system planning phases contain specified key parameters. Two of the parameters *traffic* and *coverage threshold* have a strong influence on the coverage, capacity and quality in the radio network and can thus be called "global." These parameters define the basic required configuration that is related to base station antenna heights in the radio network.

Traffic together with the available frequency band defines the number of base stations over a certain coverage area as shown already in Figure 1.6. Correspondingly, the coverage threshold defines the required number of base stations to cover the same area. By comparing these two results it can be shown whether the radio network planning is *coverage* or *capacity driven* or *limited.* When the theoretical base station quantity is studied the average base station antenna height and thus also the average base station coverage area for the radio network can be defined. The *average base station antenna height* could be the third "global" parameter because it has the strongest individual influence on the base station coverage area and furthermore the strongest influence on the frequency reuse. Hence, *traffic*, *coverage threshold* and *antenna height* must be considered during network evolution.

2.1.1 Dimensioning

Dimensioning is the first phase of the radio system planning process and its purpose is to initially draft the *radio network configuration* and *deployment strategy* for the long-term. This work could also be called a strategy of radio system planning because the aim is to define the essential radio parameter values and technologies in order to deploy the network. If the radio network is new there have to be several scenarios on how to exceed the coverage thresholds in different traffic situations. If an existing network is extended, the traffic history over the area has to be utilised to identify traffic increases during the next 1–3 years. The better the traffic forecasts the better the configuration (antenna heights and capacity) can be optimised for network evolution.

In order to study the coverage and capacity requirements for a specific area several more detailed parameters need factoring into the dimensioning, specifically knowledge of the:

- size of the covered area
- coverage threshold
- frequency band (for the radio propagation, 900/1800/2100 MHz)
- path loss between the BTS and MS (from the power budget calculations).

And for the coverage analysis and for the capacity analysis the:

- total traffic over the coverage area
- targeted maximum blocking
- frequency band (the number of the frequencies)
- frequency reuse (the maximum number of the frequencies at the base station).

The base station coverage area can be estimated by utilising the above mentioned coverage parameters, and the required number of base stations can be tuned by changing the base station antenna height to correspond to the number accordingly based on traffic requirements for the coverage area. Because traffic is increasing year after year, this analysis has to be done based on differing traffic demands, with the final network configuration and deployment strategy dependent on this long-term analysis. If this long-term analysis is not done the base station antennas will not be located correctly and the radio network configuration will not be cost-efficient due to overcapacity or load and will require continuous reconfigurations. Note that reconfigurations can not be avoided, but they can be minimised. Steadily increasing demands upon coverage (for example increased data traffic and data service) cause changes, as in coverage thresholds, and these changes have to be taken into account at the dimensioning stage because they strongly influence base station site locations.

2.1.2 Detailed radio planning

Detailed radio planning is the second phase in the radio system planning process. After defining the average base station antenna height in the dimensioning phase (based on the traffic and coverage threshold requirements) the required radio network has finally to be designed and implemented. Detailed radio planning—coverage, capacity and frequency planning—has to be done and documented. These three different planning phases are typically presented in discussion about the radio planning process. Noting the connection between coverage and capacity planning and the influence on base station configurations, the detailed planning process phases are (see also Figure 2.2):

- configuration planning
- coverage planning
- capacity and frequency planning
- parameter planning.

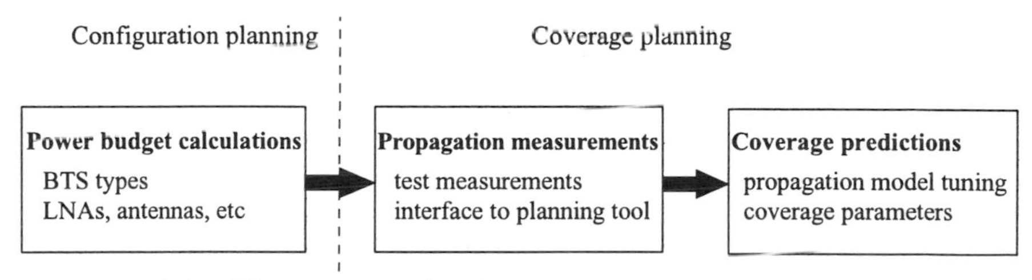

Figure 2.2. The coverage planning process.

Each of these phases are utilising dimensioning results with the aim of designing a cost-efficient and high Quality of Service radio network in practise.

2.1.2.1 Configuration planning

Configuration planning must always occur in a specified area in order that traffic and coverage thresholds are exceeded. The area definition also needs to use as constant a base station site configuration (e.g. macro sites and coverage driven configuration) as possible in the radio interface. These constant configurations make it easier to achieve the high quality radio network and they also help, for example in logistics. Configuration planning is thus needed prior to coverage and capacity planning to analyse all available coverage and capacity related hardware and software features and single-, multi band equipment, to define which features are required at which base station sites in different environments. Configuration planning analyses the capacity related system features and their influence on performance regarding the maximum number of frequencies at base stations. These capacity features and frequency assignments have a direct effect on the base station site equipment (e.g. narrowband or wideband combiners) which typically cause losses at the base station and in the antenna line. Moreover, some features like frequency hopping may also provide improvement for the path loss between the base station and mobile station. The base station site has to be configured based on both coverage and capacity requirements; the required capacity features define the capacity related base station site elements (combiners, etc.) and the required coverage or dominance area defines the need for other coverage related equipment (e.g. antenna gain, low noise amplifier (LNA), power amplifier (PA), diversity reception). The optimised base station configuration can finally be analysed by calculating the power budget for the BTS–MS connection for the uplink and downlink directions. The above mentioned capacity and coverage related equipment specifications can be taken into account in this power budget calculation and both coverage as well as capacity can be optimised. The above mentioned issues related to single band and multi band planning bring about new equipment (like multi band antennas, cables, LNAs, boosters and diplexers) and questions arising from their need at different base station sites. Multi band base station equipment—as with all base station equipment—and its performance, has to be clearly understood in detail in order to be able to decide and define the base station site configurations for different environments, during this configuration planning phase.

As a result of configuration planning the

- base station site type (*macro, micro, indoor*)
- base station antenna line (*antenna height, single-, multi band*)
- base station *coverage/dominance* (in other words *service*) *area* and *capacity*

for different areas and environments have to be defined. These three configuration areas define the total base station site configuration that is to be used for a specific area in order to maximise the radio network cost-efficiency and QOS. As only the final base station site locations may yet amend these configurations these can be clarified and confirmed in the detailed coverage planning.

2.1.2.2 Coverage planning

Configuration planning defines the base station site equipment for different environments (for coverage or capacity purposes). Coverage planning ultimately defines the radio network configuration. The aim of coverage planning is to utilise the dimensioning results (the average base station antenna height) and the configurations defined in the configuration planning (based on the power budget calculations) to minimise the number of base station sites. Thus, coverage planning has to be done over a certain area in order to be able to optimise the base station site locations and thus to utilise the base station configurations. The different phases of the coverage planning process are presented also in Figure 2.2 that shows the importance of the radio propagation measurements during radio coverage planning. Coverage planning begins with an open-minded coverage area survey which considers environmental limitations such as high buildings, hills or other obstacles. This survey indicates potential propagation problem areas and may already suggest some requirements for base station site locations. In this way this survey initially defines the critical base station site locations and suggests strategies to cover the area. Note at this stage that in certain areas it is more cost efficient to use, for example one base station site in addition to minimum base station configuration than one site less with the maximum configuration because of the environment. After this definition of the overall configuration for a certain area the propagation measurements need to be analysed—unless and only if there are measurements available about the area already. These measurements are required to tune the radio propagation prediction model which is extremely important when considering capacity and frequency planning along with the functions of the radio network.

When the coverage area and radio propagation environment have been studied and the measurements have been taken then follows the tuning of the prediction model. The measured sites are tuned and these results are utilised over the coverage planning area. This model tuning provides the accuracy for the radio coverage prediction. After prediction model tuning the candidate base station site locations are defined. The site locations are called candidate because there are no rental contracts yet for the sites and thus hypothetical site locations have to be used based on the site survey, measurement and propagation prediction results. Note that site contract negotiations and site construction take on average 4–6 months and thus the measurements and candidate site selections and the whole coverage planning process have to be started approximately twelve months before the launch of the planned sites. Moreover, the process has to be initiated with the measurements because only they can ensure accurate coverage planning.

The candidate sites are selected as if they were the final sites and they are planned as accurately as the final sites; this phase really is detailed coverage planning even though the candidates are to be confirmed. From Figure 2.2 candidate site selection goes beyond the coverage planning and is actually dealt in the same way as if it were final coverage planning. Actual coverage planning contains the radio parameter input, radio prediction model set-up and coverage area calculation for each base station. Parameter input includes all the required parameters (calculated in the power budget) to define either the optimised downlink or uplink transmission power from the base station or mobile station antenna. Radio coverage planning is traditionally calculated in planning systems (advanced software programs) in the downlink direction from the base station transmitter to the mobile station antenna by defining the

- base station transmission power
- base station equipment and antenna line losses (combiner loss, cable loss)
- base station antenna height, direction, gain, and tilting if it is used.

The radio propagation prediction model is then checked once more and tuned, and finally the base station coverage area is calculated by using the input parameters, prediction model and digital maps. Typically all these coverage planning phases are done in the radio planning tool, which provides a radio planning platform where the measurements can be imported, prediction models can be tuned and the coverage, capacity and frequency planning can be done.

The last phase in coverage planning is to define the final coverage thresholds and the coverage areas where these certain thresholds have to be exceeded. The coverage planning thresholds can be defined based on the power budget and the coverage planning margins. These thresholds also finally define the average maximum distance between two base station sites.

2.1.2.3 Capacity and frequency planning

Capacity and frequency planning are very operative and radio planning-tool based processes because the radio network configuration—as the number of frequencies used at each base station—is already decided in the dimensioning phase and the aim is only to provide the best possible result by using the required configuration. Capacity and frequency planning starts (or they have to be taken into account) when the base station sites are selected. The base station locations should be selected by trying to achieve equal base station coverage areas (and enough overlapping) and thus try to minimise interference in the radio network.

Capacity and frequency planning starts by defining the planning thresholds, which depend on the hardware and software features used in the radio network. When the thresholds have been defined the rest of the capacity and frequency planning process is planning-tool based work at the onset of the radio network deployment. When the question is about the extension of a radio network a more detailed analysis is required to understand the actual capacity needs in the radio network. This analysis again has to be done over an area and based on the traffic measurements from the radio network. The total traffic has to be gathered from the specific area and the number of frequencies have to be calculated and compared to the actual configuration.

2.1.2.4 Parameter planning

Parameter planning is actually a very short phase before the launch of the radio network because the radio network parameter values are typically fixed and because their values are based on the measurements from the other networks and thus on precedence. Typically parameters are divided into subgroups like:

- signalling
- radio resource management
- mobility management
- neighbour base station measurements
- handover and power control.

These parameters all concern and handle one type of function in the radio network. The parameter values are also quite fixed for the different environments but some small changes can be utilised, e.g. for outdoor and indoor locations. Some special cases—like the dual band radio network and the traffic distribution between the frequency layers—need more detailed radio parameter analysis. The parameters themselves and the special cases are explained later because a detailed parameter planning is more connected to the optimisation process. The radio network parameters are also very powerful because they can be used for example for the prioritisation of the base stations—traffic can be distributed to a certain base station first and then others. After parameter planning the radio network is ready for operational mode and also ready to restart the same process—dimensioning and detailed planning—from the beginning due to radio network evolution. However, some statistics (from monitoring) are required from the network and some corrections (towards optimisation) can be done with the radio network before the radio network extensions need to start.

2.1.3 Optimisation and monitoring

The actual radio system planning process contained dimensioning, coverage, capacity and frequency planning and it would be perfect if the number of mobile stations and their locations were constant and already established in the radio planning phase. Unfortunately, neither the number of mobile stations nor their location is constant and thus there is no exact information about the configuration needs of the radio network before the network is up and running and some statistical data is gathered (monitoring). This statistical data indicates the final traffic in a certain area and it shows whether the radio network has overcapacity or congestion. The monitoring results (statistical data) are also a very important input for the dimensioning phase (for the network extensions) and thus a starting point for the network evolution. The optimisation process fits the designed radio network to the actual coverage demands and traffic. The first target is to verify the coverage and to analyse whether it is good enough. Next, the traffic over a certain area is studied and if the base station coverage area is overloaded (base stations are congested) it has to be analysed whether:

- the traffic has to be balanced between the base stations or
- more frequencies have to be assigned or
- more base stations have to be implemented.

The optimisation phase is an adjustment process based on real life changes that were not taken into account in the original radio system planning, which was based on the coverage threshold requirements and traffic forecasts. Thus, both coverage and traffic verifications may trigger changes which influence back upon them. Moreover, the radio system planning process has to be repeated in the optimisation (optimisation = replanning) phase using the real information about coverage and traffic. When the actual coverage and capacity have been measured the optimisation work starts with an analysis of required base stations for traffic and continues with the coverage analysis, as in dimensioning. When the radio network configuration is defined, based on these real parameters, operational optimisation can be started. In this operational phase coverage may be improved by

- maximising the base station site configurations
- moving the base station sites.

Base station site capacities are directed at corresponding to the requirements by

- defining the actual need for frequencies at each base station location
- balancing the frequency assignments at each base station
- defining the required capacity-related software features to improve capacity.

When the base station coverage areas are satisfactory and the base station dominance areas correspond to the capacity requirements the radio network is balanced.

2.1.4 Radio system planning process documentation

The different radio system planning phases (as introduced) create a radio system planning process where there are

- a strategic system planning part (dimensioning)
- an operative system planning part (detailed radio planning)
- a system planning monitoring part (optimisation and monitoring).

All of which can be managed and controlled by having good input and output documents. These input and output documents are critical in the whole radio planning process because the common planning criteria (needed in order to achieve constant quality in the radio network) can only be applied if it is documented exactly (input documentation). Moreover, knowledge of the radio network configuration can only be upgraded if the output documentation is accurately and correctly organised. Thus, exact documentation is needed to control the radio system planning process, to guarantee the high quality and cost efficient radio network infrastructure and to manage the radio network in the long-term.

Figure 2.3 introduces the radio system planning folder which is a documentation database whose function is to guide the radio system planning process. The planning folder can be implemented in the planning system or it can be managed by a separate software application or the common directory trees can be used (as in Figure 2.3).The planning folder needs to be located in a common area (for example in a shared drive on an operator's local network) such that the whole radio planning organisation has open access to it.

The planning folder includes input documents (open reading access) and output documents (open reading and writing access). The input documents are in the planning and R&D (research and development) directories and the output documents are in the site folder directory (highlighted). The input documents contain the information that has to be controlled and accepted by a planning manager and the output documents contain data that is related to the radio network configuration.

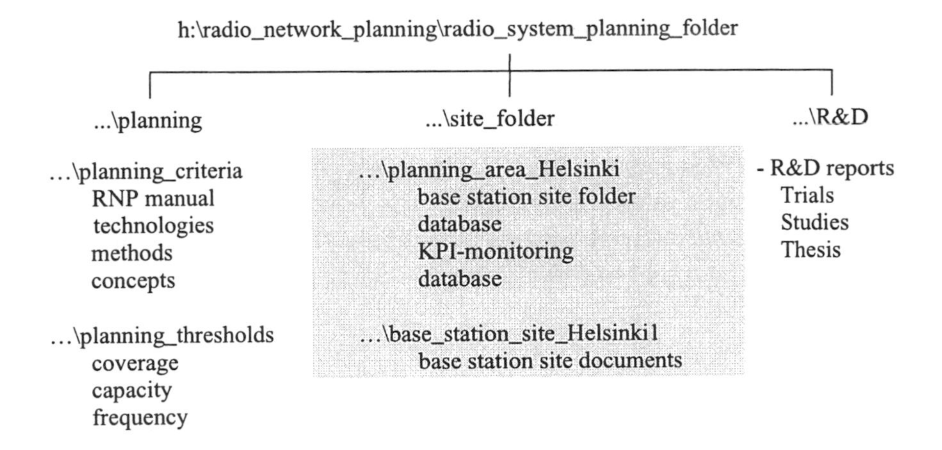

Figure 2.3. The radio system planning folder.

Input documents

The input documents include the data in the planning and R&D directories. The planning directory has two subdirectories called planning_criteria and planning_thresholds. The planning_criteria directory contains the technical information about the technologies, methods and concepts that are used in radio system planning and deployment. These items should be documented in the radio network supplier based *radio network planning (RNP) manual* that defines and specifies the radio system planning criteria. The radio network planning manual should be as complete a description as possible regarding radio system planning because it is the only guarantee that common rules are used in radio system planning throughout radio network evolution. Additionally it can be efficiently used for training and as an introduction to new radio planners.

The second input subdirectory is planning_thresholds; based on its contents. As this suggests, the directory contains the final radio system planning thresholds such as:

- planning strategy (dual band, GPRS, EDGE, UMTS)
- planning platform (tools, interfaces, etc.)
- the available base station site equipment and their performance
- coverage thresholds

- frequency band (available frequencies, maximum number of transceivers at the base stations, etc.)
- quality criteria (maximum blocking, drop calls, handover failures, etc.)
- the features used and functions in the radio network (frequency hopping, etc.)
- default radio parameter settings (e.g. dual band parameters)
- monitoring principles (key performance indicators (KPI)).

These documents are the final guidelines for the radio planning engineers to follow and they also define the targeted QOS in the radio network. It is also critical to have a good document about the planning platform in order to define exactly all of the planning tools and interfaces for the different radio system planning phases.

The last input (and also output when accepted by the planning manager) directory is R&D, which is required to provide a common forum for all the trial and test reports and studies. This directory is a database for all technical documents which are related to radio system planning and which are not defined in the planning_criteria and planning_thresholds directories. These documents are typically reports which support the planning criteria and which can be used to give a wider understanding of the radio planning topics.

Output documents

The output documents are in the site_folder directory which contains subdirectories for each radio planning area, which in turn contain two databases (base station site folder and KPI-monitoring) and subdirectories for each base station sites. These documents have information required about the radio network configuration and its performance at each moment.

The documentation of the radio network configuration is based on the base station site folder (database) and on the base station site directories. Each base station site directory contains a document called base station site data which contains all the required information to build a certain base station site. This base station site data document is used to approve the radio planning for each base station site. The base station site data document is also copied to the base station site folder in order to be able to monitor the whole radio network infrastructure (the site folder should include all the base station site equipment and parameters). Note that the same information that is in the base station site data and in the base station site folder is also in the radio planning system in order to draw up a radio plan.

Thus, the same data is documented three times for different purposes:

- data in the planning system is for the actual planning
- the base station site data is for the acceptance of the radio planning (typically printed and signed by a radio planner)
- the base station site folder is for the monitoring.

The other type of output documentation is the radio network monitoring that can be done by using key performance indicators. The importance of these KPI values is to build up a history, e.g. about traffic increase in different geographical areas. By recording the KPI values, say at least once a month, the long-term development and evolution that is required from all radio planning areas builds up.

The input and defined output documents are necessary for successful radio system planning in the operational phase but of course other documents are needed e.g. from dimensioning. However, these documents in the site folder control the radio system planning process and automatically take care of the quality. The different radio system planning phases and output documents (or documentation) are thus linked such that the base station site data is always documented first when the coverage plan is made (from site search to site selection). When the radio plan for the selected site is accepted, implementation can be started. When the site has been built the radio parameters have to be assigned and, for example base station identifications, location areas and designated frequencies have to be added to the base station site document. When the site is finally in operation the site folder has to be upgraded and the monitoring of the site started. These output documentation phases happen time after time during an evolution of the radio network and some changes are needed for both output and input documents due to radio network extensions and the introduction of new technologies.

2.2 Conclusions

The radio system planning process is carried out in phases and has to be carefully documented in order to maximise the cost efficient radio network configuration. The critical planning phases for long-term success (see Table 2.1) are dimensioning and monitoring; as they define the radio network deployment strategy. The other planning phases are driven more by technical details which, of course, have a great effect on the final radio network configuration. However, the whole planning process can be optimised when the documentation is done correctly and the input and output documents guide the radio system planning procedures.

Table 2.1. The highlights of the radio system planning process.

Subject	Findings
Dimensioning	–Global parameters: traffic, coverage threshold and base station antenna height
Detailed radio planning	–Practical issues
Configuration management	–Optimised features for different radio network evolution phases
Coverage planning	–Accurate coverage predictions
Capacity and frequency planning	–Accurate coverage predictions are a *must*
Parameter planning	–Radio network functions
Optimisation	–Adjustment of the radio network
Radio system planning process	–Importance of documentation –Site folder

Chapter 3

CONFIGURATION PLANNING AND POWER BUDGET

3. CONFIGURATION PLANNING AND POWER BUDGET

3.1 General power budget

Base station site configuration defines the maximum allowed path loss and includes the *base station equipment* and *antenna line configurations* as illustrated in Figure 3.1.

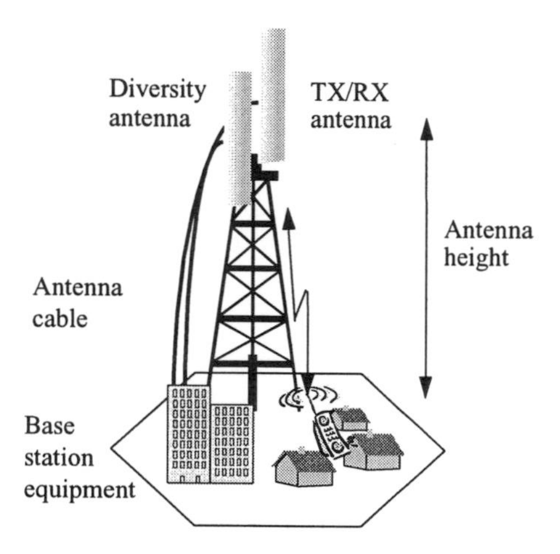

Figure 3.1. Base station site configuration includes the equipment and antenna line configurations.

Base station equipment and antenna line configurations can be designed for different purposes by solving the path loss in the power budget. They also include both general and accessory elements and parameters. The equipment configuration elements and parameters are

- *sensitivity (uplink)*
- *transmitting power (downlink)*
- *combiner loss (downlink)*

and the antenna line configuration elements and parameters are

- *receiving (RX) and transmitting (TX) antenna gains (uplink, downlink)*
- *cable loss (uplink, downlink).*

These elements are needed to build up a basic base station that transmits and receives a radio signal.

The basic configuration and power budget can be improved by introducing features and accessory elements which are

- *frequency hopping (uplink, downlink)*

for equipment configuration, and

- *diversity reception (uplink)*
- *low noise amplifier (LNA, uplink)*
- *booster or power amplifier (downlink)*
- *duplex filter (uplink, downlink)*
- *diplex filter (uplink, downlink)*

for the antenna line configuration. The technical performance of these elements is manufacture related. These elements are also configuration related. The equipment configuration therefore has a link to the antenna line configuration and vice versa. Note also that the performance of certain parameters like diversity reception and frequency hopping are strongly related to the radio propagation environment and thus urban, suburban and rural areas require different corrections in their power budgets.

The technical parameters of a mobile station (MS) have to be defined before the power budget calculations can be done. The mobile station parameters

- *transmitting power (uplink)*
- *sensitivity (downlink)*
- *transmitting (TX) and receiving (RX) antenna gains (uplink, downlink)*
- *cable loss (uplink, downlink)*

have to be included in the power budget. Remember that mobile stations are standardised consumer products and thus *typical technical values have to be known because they cannot be configured separately.*

In order to calculate the maximum allowed path loss between the base station and the mobile station and to select the optimised base station site configuration all these parameters have to be collated in the power budget (Figure 3.2).

UPLINK

Mobile station	Unit	Value	
RF power	dBm	33	A
Cable loss	dB	0	B
TX antenna gain	dBi	0	C
Peak EIRP	dBm	33	D = A – B + C
Base station	**Unit**	**Value**	
RX antenna gain	dBi	16	E
Cable loss	dB	3	F
BTS sensitivity	dBm	– 106	G
Minimum reception level	dBm	– 119	H = – E + F + G
Isotropic path loss	dB	152	I = D – H

DOWNLINK

Base station	Unit	Value	
RF power	dBm	45.0	A
Combiner loss	dB	2.5	B
Cable loss	dB	3.0	C
TX antenna gain	dBi	16.0	D
Peak EIRP	dBm	55.5	E = A – B – C + D
Mobile station	**Unit**	**Value**	
RX antenna gain	dBi	0.0	F
Cable loss	dB	0.0	G
MS sensitivity	dBm	– 105.0	H
Minimum reception level	dBm	– 105.0	I = – F + G + H
Isotropic path loss	dB	160.5	J = E – I

Figure 3.2. General power budget for the GSM900 system without accessory equipment.

The two main areas of Figure 3.2 represent the uplink (MS transmitting and BTS receiving) and downlink (BTS transmitting and MS receiving) directions. The aim is to balance and maximise the allowed isotropic path loss by maximising the transmitting power of the base station.

All elements thus have a direct effect on the base station transmitting power. The parameter values for the mobile stations have to be constant and typically the value 0 dB is used for antenna gain and cable loss if no external antennas and cables are used. As equipment configuration parameters are fixed when the base station supplier is selected it is the base station antenna line parameters that have the strongest effect on the final path loss.

Figure 3.2 shows the allowed path loss in the uplink direction is 8.5 dB less than in the downlink when the base station transmitting power is 30 W (= 45 dBm). This 8.5 dB difference means that the base station coverage area in the downlink direction is greater *than in the uplink direction.* The power budget can be balanced either by reducing the BTS transmitting power (thus reducing downlink coverage) or by using accessory elements for the uplink (while retaining downlink coverage).

3.1.1 Base station and mobile station transmission power

GSM specification 05.05 says that the BTS transmission RF peak power can, for example, be 20–40 W (TRX power class 5) at the 900 MHz frequency band. The GSM manufacturers typically have a few macro BTS types with maximum power (typically in the power class 5) and also some special BTS types for transmitting lower peak powers. As different suppliers have different BTS types with transmission power from 1.0 to 50 W, the nominal peak power of each BTS type has to be carefully checked. Note that transmission peak powers can vary between the frequency bands 900/1800/1900 MHz.

There are five different mobile classes for the GSM900 system (see Table 3.1). Classes 1–3 are vehicle mounted and are something of a rarity, whereas handhelds, the mobile classes 4 and 5, are the mainstay of the GSM system, specifically the handheld with 2 W transmission peak power. Handhelds of 0.8 W have been discussed but they are not widely supported at present.

Table 3.1. GSM mobile station transmission power at 900 MHz frequency band.[1]

MS class	TX power (W)
1	—
2	8.0
3	5.0
4	2.0
5	0.8

GSM specification 05.05 states that mobile station class 4 peak power at 900 MHz has to be 2 W ± 2 dB. Manufacturers utilised this ± 2 dB margin in the early 1990s and typically delivered mobile stations with around 1.5 W transmission peak power. This lower transmission power saved the mobile battery, one of the key topics when performance was discussed. In the late 1990s mobile battery development had improved to the extent that it ceased to be an issue and suppliers started to develop mobiles closer to the 2 W peak transmission power criteria. Competition between mobile suppliers has led to improved transmission power (and therefore coverage area) at the expense of battery life. In the near future we may expect to see nominal power transmitted at 2 W + 2 dB.

3.1.2 Base station and mobile station sensitivities

GSM specification 05.05 defines the base-line minimum reception levels (without diversity reception), as – 104 dBm and – 102 dBm (900 MHz) for the BTS and MS, respectively. However, manufacturers have started to improve base station sensitivity levels and to indicate values lower than – 104 dBm, even as low as – 108 dBm. The environment for these values is not revealed, nor is whether diversity reception was applied. When sensitivity levels are discussed, the environment and the number of reception branches should be indicated so as to confirm the base-line.

3.1.2.1 Types of environment

In GSM specification 05.05 there are five different area types defined for the GSM900 system:

- *static,*
- *typical urban 50 km/h* *(TU50)*
- *hilly terrain 100 km/h* *(HT100)*
- *rural 250 km/h* *(RA250).*

In these areas the radio propagation fading conditions are different and thus base station sensitivity also differs. Therefore, accurate sensitivity values must be used in different environments or else the poorest value must be used over the whole network in order to avoid pockets of quality reduction.

3.1.2.2 Base station sensitivity and diversity reception

Diversity reception is also sometimes connected to BTS sensitivity and leads to discussion about *receiving system sensitivity*. Depending on the diversity technique and environment the typical two branch diversity reception gain is 3–6 dB. Therefore, BTS sensitivity is also 3–6 dB better if the diversity reception is included in the sensitivity values. Sensitivity measurements are typically simulated in an optimum condition using *zero correlation* for the different reception branches. In reality the signal correlations of the different receiving branches are not even close to the zero and the *signal levels are not equal.*

The actual parameter ranges have to be taken into account when BTS sensitivity and diversity reception is determined.

3.1.2.3 Mobile station sensitivity

Mobile station sensitivity covers the same parameters as the base station but diversity reception is not typically used at the mobile station receiving end. Mobile station suppliers have also improved sensitivity: the value of – 104 dBm, or better, has been frequently recorded. However, nominal values of different mobile station types have to be measured in different environments. Typically values from – 102 dBm to – 105 dBm have been safely used in radio planning.

3.1.3 Combiner and receiving multicoupler losses

Combiners in the downlink direction and receiving multicouplers in the uplink direction are needed in the base stations if more than one tranceivers are assigned to the same antenna line. The combiners merge the frequencies to the same antenna line in the downlink direction and simultaneously cause attenuation. The receiving multicouplers separate the frequencies in the uplink direction but their attenuation is negligible. Combiner attenuation is related to the combiners' performance. Combiners can be narrowband and thus tuned to a certain frequency band so the attenuation can be minimised. Wideband combiners (of higher attenuation) are typically needed if frequency hopping is used and frequencies are far away from each other. Table 3.2 gives an idea of the different combiner types and attenuations. Wideband combiners are, of course, meant for capacity areas where a maximum number of tranceivers are required over each coverage area. In these capacity areas many base stations are needed, coverage is limited and thus minor higher combiner loss does not cause any serious difficulty (however indoor coverage is reduced). Correspondingly, frequency band selective combiners are used in the areas where maximum coverage is required and thus combiner loss is also minimised. Combiner loss can be minimised when it is by-passed but then only one frequency can be assigned to the antenna line (a new antenna line is required if capacity is needed in the future for a second frequency).

Table 3.2. Typical combiner losses in the different BTS configurations.

Combiner type	Loss (dB)
by-passed	2–3
narrowband	3–5
wideband	5–7

3.1.4 Base station antennas

"Antenna" typically describes an entire radiating element, connected via a line to the base station equipment (see Figure 3.1). Specifically "antenna" can be a short piece of metal wire (a wire antenna, as in a dipole antenna) or a metal plate (patch antenna). These examples are based only on one element and therefore "antenna" is often combined with some other word, as in "antenna element." When two or more of these elements are connected they form an "antenna array." Antenna arrays achieve direction (also called gain) for the radiation.

Base station antennas, depending on their application, comprise either one antenna element (small size, low gain and multi band antennas without diversity) for indoor applications, or an antenna array (high gain and directional) for macro cell applications.

The essential base station antenna parameters are:

- Gain (low/medium/high)
- Beamwidth (horizontal and vertical)
- Size
- Polarisation
- Diversity technique
- Frequency band
- Tilting properties.

Environmental aspects together with coverage and capacity related issues determine the system level requirements for the whole base station antenna line, including the antenna itself.

Antenna gain, size and *beamwidth* need to be considered simultaneously as they have strong links to each other. Macro base stations (antennas above the average rooftop level) in urban and rural areas typically have to cover as large an area as possible. A high gain base station antenna (e.g., 18 dBi at 900 MHz) is one of the best means of achieving high coverage in this environment. However as the height of a high gain antenna is between 1500–2500 mm it will require careful planning—awareness of visual impact—and sensitive installation. The *antenna beamwidth* parameter provides some assistance when flexibility is needed in the antenna gain and size.[2] The typical pattern for the antenna horizontal half power radiation, for macro base stations, is 60–120°, achieving maximum *gain* and *coverage area* but also maximum *capacity* in the radio network. Note, base station antenna strongly influences capacity (see Chapter 6). A small variation in the horizontal beamwidth can be achieved by the antenna size and vertical beamwidth. Commercial high gain antennas (e.g., 18 dBi at 900 MHz) typically have a horizontal beamwidth of 65° and vertical beamwidth of 7–9°. This vertical beamwidth is at a bare minimum, with no reserve. In macro base stations a narrow vertical beamwidth is an advantage, as transmission power can be directed to a certain area (known as tilting, also see Chapter 6).

Moreover, when base station antennas are installed below the rooftop level—on outdoor walls in micro base stations and on inside walls in indoor base stations—antenna size is the most critical parameter. In outdoor locations antenna gain should be maximized to increase indoor coverage and therefore a narrow horizontal beamwidth is not too critical as buildings neighbouring the antenna will alter it anyway. The vertical beamwidth can, in an outdoor micro base station, be quite wide (up to 45°) in order to maximise coverage to the upper floors of high-rise buildings. Vertical beamwidth should be selected by taking into account the need for antenna gain for indoor coverage and building heights (constant 4–5 floor buildings or tower blocks). Correspondingly, in indoor locations the base station antenna should be, of course, very small (one element antenna). Thus, antenna gain can also be quite small (the coverage is needed only for indoor locations), the horizontal beamwidth can be omnidirectional or direction-specific depending on the shape of the building and the vertical beamwidth can be flexible.

Vertical *polarisation* is primarily applied in mobile communication systems as it has better propagation properties in urban outdoor locations than horizontal polarisation.[3] Vertical polarisation has lower losses than the horizontal due to diffraction and reflections caused by buildings and other obstacles. There are also many more vertical surfaces for reflections than horizontal ones. Therefore, vertical polarisation is primarily used and recommended for outdoor and indoor locations. The exception to this recommendation is a tunnel environment, where polarisation is selected mainly based on the tunnel shape.[4]

Diversity reception is also utilised widely in mobile communication systems in order to improve the power budget and thus to improve the quality at the base (and/or mobile) station receiving end. Traditionally space diversity was used with two base station antennas which were physically separated (creating a large antenna configuration). Nowadays polarisation and other diversity techniques are more commonly used and studied because they require no physical separation and thus only one base station antenna location is needed. These different diversity reception techniques improve the macro and micro coverage areas. In indoor base stations diversity is not required as the systems typically provide a high enough received field strength level to achieve good quality. Diversity is also very expensive for indoor systems as the cabling should be doubled.

Antennas are designed to function in a single *frequency* band, dual band or multi band. Single band antennas are, of course, easier to manufacture whereas dual/multi band antennas are produced for specific purposes. The need for single or multi band antennas can be assessed by understanding the radio propagation in different environments. In urban and rural macro base stations, the coverage area differs between 450 MHz (GSM), 900 MHz (GSM), 1800 MHz (GSM) and 2100 MHz (UMTS) frequency bands and thus, the same antenna location is not necessarily the optimum solution to deploy the multi band radio network. Therefore, single band antennas are primarily used in macro base stations and multi band antennas are utilised when necessary. Multi band antennas provide an economical solution in macro and micro base stations and in indoor applications as only one antenna and cable has to be installed.

The last antenna related parameter is *tilting*, required in order to decrease interference in high capacity areas where base station coverage has to be limited and focused. Tilting directs the vertical main beam, typically between 5–20° (depending on the vertical beamwidth) from the horizon towards the earth's surface. Tilting can be done manually but the use of electrical tilting is usually recommended. As tilting is typically necessary for urban macro base stations (in a high capacity area) it is critical to have quite a narrow vertical beamwidth in the base stations in order to achieve maximum flexibility in tilting (if the vertical beamwidth is close to 15°, a huge tilt is required to obtain any results). Downtilting is a standard solution but in some special cases uptilting may be required, as in hilly areas. In micro base stations and indoor applications tilting is usually not needed (as a wide beamwidth is recommended here).

In addition to the system level requirements at 900 MHz (see Table 3.3), other frequency bands' data are similar, with the exception of antenna gains (that can vary).

Table 3.3. Base station antenna system level requirements at 900 MHz.

	Urban macro	Rural macro	Micro	Indoor
Gain (dBi)	12–18	16–18	7	7
Beamwidth horizontal	65–80	65–80	65–90	65–360
Beamwidth vertical	7–10	7–10	< 45	Not critical
Diversity	Polarization	Polarization	Polarization	No
Polarization	Vertical	Vertical	Vertical	Vertical / Horizontal*
Tilting	Yes	Yes / No	No	No
Frequency band	Single / Dual	Single / Dual	Multi	Multi
Size	Large / Medium	Large / Medium	Small	Small

* Tunnel applications need different polarization to maximize the radio propagation.

3.1.5 Base station antenna installations

When a base station antenna location is selected various issues have to be taken into account:

- Height
- Obstacles next to the antenna
- Backlobe
- Coupling.

These factors have a direct influence on radio propagation and thus on the quality of the radio network. *Antenna height* is a very critical parameter as it has the strongest influence on the coverage prediction results which furthermore influence frequency planning. The antenna height also influences the coverage prediction models which typically function either in the macrocellular or in the microcellular propagation environment. Moreover, the area around the antenna should be free of obstacles (at least for 50–100 m) in order to be able to use the radio propagation prediction models and to trust the prediction results. The actual antenna height should always be measured and documented carefully. All of a base stations antennas should also be installed at a consistent height, as far as possible, as this influences the radio propagation and also has an effect on the frequency reuse and therefore on the maximum capacity in the radio network (see Chapters 6 and 7).

In planning, *antenna backlobes* can be avoided by installing antennas to exterior building walls and by using tilting, especially in urban areas where the base stations interfere with each other. *Antenna coupling* should also be considered if polarisation diversity or duplex filters are not applied and more than one base station antenna is used (for example, space diversity in rural areas). Transmitting and receiving antennas cannot be installed too close to each other in order to get sufficient attenuation. It is very important to get updated installation instructions from the base station antenna suppliers because these installation instructions vary significantly depending on the antenna elements (dipoles or patches) and the separation may vary, e.g. from 0.1 m to 1.5 m in the horizontal direction (less typical in the vertical direction) in order to achieve for example 30 dB isolation. Polarisation diversity and duplex filters are typically manufactured with an isolation of 30 dB between the transmitting and receiving branches that is sufficient for GSM.

3.1.6 Mobile station antennas

Mobile station antennas are typically omnidirectional or 180° (not radiating towards a head) in the horizontal plane and very wide in the vertical plane. Antenna gain is assumed to be 0 dBi because mobile antennas are very small and because the power budget has to be calculated by taking into account all antennas in the field. If some external antennas are used, as in vehicle installations, antenna gain could be included if needed.

3.1.7 Base station cables and connectors

The base station antenna line also includes cables and connectors between the antenna and base station. Cables can be thin (jumpers) with a high attenuation loss, or thick (1 5/8") with quite a low attenuation. The attenuation of typical cables 1/2" or 7/8" at 100 m (at 900 MHz and 1800 MHz) are presented in Table 3.4, and Table 3.5 presents losses more precisely at 900 MHz.

Table 3.4. Attenuations of the 1/2", 7/8", 1 5/8" and jumper cable types of 100 m at 900 MHz and 1800 MHz.

Type	900 MHz (dB)	1800 MHz (dB)
1/2"	7.7	12.0
7/8"	4.0	7.0
1 5/8"	3.0	5.0
Jumper	0.3	0.5

Table 3.5. Attenuations of the 1/2" and 7/8" cable types with jumpers at 900 MHz.

Length (m)	1/2"	7/8"	7/8" + jumper	Length (m)	1/2"	7/8"	7/8" + jumper
5	0.4	0.2	0.8	80	6.0	3.2	3.8
10	0.8	0.4	1.0	85	6.4	3.4	4.0
15	1.1	0.6	1.2	90	6.8	3.6	4.2
20	1.5	0.8	1.4	95	7.1	3.8	4.4
25	1.9	1.0	1.6	100	7.5	4.0	4.6
30	2.3	1.2	1.8	105	7.9	4.2	4.8
35	2.6	1.4	2.0	110	8.3	4.4	5.0
40	3.0	1.6	2.2	115	8.6	4.6	5.2
45	3.4	1.8	2.4	120	9.0	4.8	5.4
50	3.8	2.0	2.6	125	9.4	5.0	5.6
55	4.1	2.2	2.8	130	9.8	5.2	5.8
60	4.5	2.4	3.0	135	10.1	5.4	6.0
65	4.9	2.6	3.2	140	10.5	5.6	6.2
70	5.3	2.8	3.4	145	10.9	5.8	6.4
75	5.6	3.0	3.6	150	11.3	6.0	6.6

The number of cables can be reduced by connecting the transmitting and receiving branches by using a duplex filter and in multi band cases the frequency bands can be connected by using diplexers. Both duplexers and diplexers cause less than 1.0 dB loss, including connector loss in both directions.

3.1.8 Mobile station cables and connectors

Mobile station cable and connector losses are typically assumed to be 0 dB. If some external antennas and cables are used, as in vehicle installations, some cable loss could be included in the power budget if needed.

3.2 Power budget and accessory elements

3.2.1 Power budget balance

The power budget in Figure 3.2 was not balanced and permitted accessory elements are needed for the uplink direction to increase the path loss. When the low noise amplifier (LNA) and diversity reception are added to the power budget the result can be seen in Figure 3.3. The depolarisation loss is also included in the power budget in order to take into account the possibility to use polarisation diversity. Figure 3.3 shows that the uplink direction is a little better than the downlink direction if the values 4 dB and 5 dB are used for the diversity reception and low noise amplifier, respectively. It has to be remembered that diversity gain depends on the radio propagation environment and base station antenna configuration.

UPLINK

Mobile station	**Unit**	**Value**	
RF power	dBm	33	A
Cable loss	dB	0	B
TX antenna gain	dBi	0	C
Peak EIRP	dBm	33	D = A – B + C
Base station	**Unit**	**Value**	
RX antenna gain	dBi	16	E
Diversity reception	dB	4	F
Depolarization loss	dB	0	G
Low noise amplifier	dB	5	H
Cable loss	dB	3	I
BTS sensitivity	dBm	– 106	J
Minimum reception level	dBm	– 128	K = – E – F + G – H + I + J
Isotropic path loss	dB	161	L = D – K

DOWNLINK

Base station	**Unit**	**Value**	
RF power	dBm	45.0	A
Combiner loss	dB	2.5	B
Cable loss	dB	3.0	C
TX antenna gain	dBi	16.0	D
Peak EIRP	dBm	55.5	E = A – B – C + D
Mobile station	**Unit**	**Value**	
RX antenna gain	dBi	0.0	F
Cable loss	dB	0.0	G
MS sensitivity	dBm	– 105.0	H
Minimum reception level	dBm	– 105.0	I = – F +G + H
Isotropic path loss	dB	160.5	J = E – I

Figure 3.3. Accessory elements for the power budget of the GSM900 system.

3.2.1.1 Low noise amplifiers

Cable losses may cause limitations in the power budgets in the uplink direction, especially if cables longer than 75 m are used (as can be seen in Table 3.5). This is common in rural areas where 100–120 m masts are needed to maximise the coverage area. The uplink direction can be improved by introducing a low noise amplifier (LNA, also called a mast head pre-amplifier MHPA, or tower-mounted pre-amplifier TMPA, or tower top amplifier TTA) next to the receiving antenna. The LNA has to have a low noise figure in order to improve the received field strength level at the base station receiving end. The improvement at reception can be studied by using the formula for calculating the noise figures for the whole antenna line from the receiving antenna to the BTS receiver (see Figure 3.4).

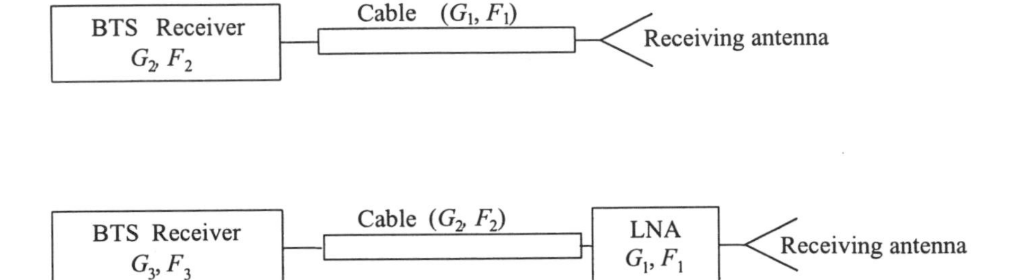

Figure 3.4. An antenna line at the base station receiving end without LNA and with LNA.

The formula for cascade amplifiers can be written as:[5]

$$F = F_1 + \frac{F_2 - 1}{G_1} + \frac{F_3 - 1}{G_1 G_2} + \ldots \qquad \textit{Equation 3.1}$$

where F_1 is the LNA's noise figure and $F_2, F_3, \ldots, F_N$ are the noise figures of the other antenna line elements and correspondingly G_1 is the LNA's gain and $G_2, G_3, \ldots, G_N$ are the gains of the other antenna line elements. It can be seen that the LNA's noise figure and gain are important parameters when analysing the LNA's performance. Regarding the optimised values for noise figure and gain, Equation 3.1 clearly shows that the lower the LNA's noise figure the lower the system's noise figure. Thus, the first target is to have as low a noise figure for the amplifier as possible: typical values are around 1.5 dB, which is almost the minimum value. If, for example, an extra cooling system is introduced it is even possible to have values below 1.0 dB. When the LNA's and BTS's noise figures are known it is also possible to calculate the maximum improvement that LNA can provide for the antenna line. This improvement can be written as

max improvement = cable loss + (BTS noise figure – LNA noise figure)

where max improvement gives the maximal improvement that can be gained and depends, not surprisingly, on the LNA's gain.

We can calculate the optimised gain for the LNA:

- Assume LNA has the noise figure = 1.5 dB and the gain is 14 dB or 21 dB;
- additionally, the BTS noise figure is 5.0 dB,
- thus, the maximum extra gain after the cable loss reduction is 5.0 – 1.5 = 3.5 dB.

Equation 3.1 gives the improvements of the system noise figure (the noise figure for the whole antenna line) and the results are presented in Table 3.6.

Table 3.6. System noise figure improvements when the 14 dB or 21 dB LNA gain is applied.

Cable loss (dB)	LNA 14 dB (dB)	Extra (dB)	LNA 21 dB (dB)	Extra (dB) 14–21 dB
2	5.0	3.0	5.4	0.4
3	5.9	2.9	6.4	0.5
4	6.7	2.7	7.3	0.6
5	7.5	2.5	8.3	0.8

Table 3.6 shows the benefit of the different LNA types when the BTS noise figure is 5 dB. It can be noted that the gain of 14 dB is enough to almost achieve the maximum advantage when typical cable loss 2–4 dB is considered. The LNA of 21 dB gain and 1.5 dB noise figure gives more than 1.0 dB extra gain if the cable loss is more than 5.0 dB. However, in radio planning such a high cable loss should be avoided and in extraordinary situations (as these) a downlink booster is often required.

The LNA also has an effect on the performance of diversity reception because it depends on the received signal levels of the main and diversity branches. The signals of the main and diversity branches should be received at an equal level and thus the LNA should be installed to both receiving branches (if used with the diversity reception).

3.2.1.2 Diversity reception at the base station receiving end

Diversity or multiple reception is required at the base station end because of the signal level, location probability and grade-of-service (GOS) deterioration in a radio propagation channel due to fast fading (see Figure 1.11). The fast fading contains fading dips of up to 30 dB corresponding to Rayleigh distribution in non-line-of-sight (NLOS) signal paths.[6-7] Moreover, the received signal level is Rice distributed, in a line-of-sight (LOS) path. The fast fading dips can be reduced by receiving multiple uncorrelated signals (diversity reception) at the same or delayed time at the base (and/or mobile) station receiving end. When the better received signal level is chosen, the average received signal level (see Figure 1.11), location probability and base station service area can be increased.

Different diversity reception schemes are based on a technique to provide uncorrelated signals. Traditionally, the base station receiving antenna arrays are spatially separated in the horizontal, vertical or compound direction and this technique is called a *space diversity scheme*. The base station antenna array separation causes different multipath lengths between the mobile station and base station and thus it provides a phase difference at the base station receiving end to obtain uncorrelated signals. These uncorrelated signals can also be provided by separating only two antenna elements. In the *polarisation diversity technique* the signals are received by using two orthogonal polarisations (e.g. horizontal and vertical, or ± 45° slanted polarisations) at reception. *Beam diversity techniques*—jitter, switched pattern and multi state phase diversities—are based on the modified antenna radiation patterns. Moreover, the uncorrelated signals can also be provided by using different *frequencies* or by receiving the signals *at different times* (delaying the diversity reception). It can be concluded that each diversity scheme contains a different technique to provide uncorrelated signals;

- receiving antenna array separation
- orthogonal polarisations
- individual radiation patterns
- frequency difference
- time delay.

In addition, all these diversity techniques can be applied either at the base station or at the mobile station end. However, only space, polarisation and beam diversities at the base station receiving end are the primarily diversity schemes because these diversity techniques have *no primary system requirements* (minimum frequency band, maximum time delay at the base station end).

3.2.1.3 Diversity reception and radio propagation environment

The performance, that is, the improvement of the signal (or actually bit-error-rate at a certain field strength level) of different diversity techniques, depends finally on the environment (three major classes)—urban, suburban and rural—and two special classes—microcellular and indoor. The different propagation environments have different propagation characteristics thus the performance of the diversity reception differs. Moreover, the performance of the diversity reception is actually equivalent in all environments if uncorrelated signals can be provided. The problem is that a large antenna configuration is required in order to obtain uncorrelated signals, for example, in the space diversity in rural areas.

3.2.1.4 Performance of deployed diversity techniques

Space diversity schemes

The performance of the space diversity scheme depends strongly on the separation of the receiving antennas in the different environments. If the separation is sufficient uncorrelated signals can be provided and the diversity gain of 5 dB can be obtained in all environments. The required separation is very small (1–4 λ) in indoor and microcellular environments,[8] and thus a very compact antenna configuration can be utilised to provide 5 dB gain. In urban and suburban macro base stations separation of 10–15 λ is enough to provide the equal gain.[9] The problem environment is the rural, where more than 20 λ separation is needed, which is too large an antenna configuration. However, if two space diversity antennas are used next to each other in the rural environment, the "combining gain" of 3 dB can be guaranteed if the maximal ratio combining is used. The combining gain means that the signal is received twice at the same phase and at the same level. This is the case in the space diversity scheme if the same antennas are used at both receiving branches and these antennas are close to each other.

Polarisation diversity schemes

Cross polarisation discrimination (XPD) depends strongly on the environment and determines the final polarisation diversity gain as the signal cross-correlation of the different polarisation diversity schemes are always clearly less than 0.5 in LOS and NLOS environments.[8] Low

XPD values (< 7 dB) can be achieved in the NLOS path in urban, suburban, microcellular and indoor environments where there are multiple reflections from different surfaces. In the rural area and in LOS paths one polarisation easily dominates and XPD values are 10–15 dB, thus reducing the diversity gain. The maximum diversity gains of 5–6 dB at 900 MHz and 1800 MHz have been measured in urban, suburban, microcellular and indoor areas when ± 45° slanted polarisations are used at the base station end.[8] This polarisation diversity scheme performs equally with the horizontal space diversity scheme of 20 λ separation between the receiving antennas. When the horizontal and vertical polarisations are applied at the base station end at 900 MHz and 1800 MHz, the diversity gain is approximately 1 dB worse than the horizontal space diversity of 20 λ separation.[8] Thus, a good performance in NLOS built-up areas can be achieved by applying ± 45° polarisations at reception. The polarisation diversity gain is also approximately 1 dB lower at 900 MHz compared with the results at 1800 MHz. The results of the diversity measurement campaigns,[8] are gathered in Table 3.7. Finally, it has to be remembered that a depolarisation loss in the uplink direction is included in the power budget in Figure 3.3. This depolarisation loss refers to the attenuation whether the vertical or slanted linear polarisation is used.

Table 3.7. Polarisation diversity gains in the different environments when a GSM quality class 3 is exceeded with 90 percent signal reliability.

Area type	Diversity gain with ± 45° polarizations, MS (0°) / MS (45°)	Compared with horizontal diversity	Diversity gain with HV* polarizations, MS (0°) / MS (45°)	Compared with horizontal diversity
Indoor	6 dB / 6 dB	0 dB / 0 dB	5 dB / 6 dB	– 1 dB / 0 dB
Urban	6 dB / 6 dB	0 dB / 0 dB	5 dB / 5 dB	– 1 dB / – 1 dB
Semi-urban microcellular	4 dB / 5 dB	0 dB / + 1 dB	3 dB / 5 dB	– 1 dB / + 1 dB
Light suburban	3 dB / 3 dB	0 dB / 0 dB	3 dB / 2 dB	0 dB / – 1 dB

* HV = horizontal/vertical

3.2.2 Power budget maximizing

The power budget in Figure 3.3 was balanced by using a low noise amplifier (LNA) and diversity reception. Now we turn to maximise the allowed path loss and include the required margins to the power budget. Figure 3.5 shows an example of the maximum power budget for the

GSM900 system when all available equipment are utilised. The path loss in Figure 3.5 is 11 dB better compared to the path loss in the basic power budget in Figure 3.2. The path loss is improved in Figure 3.5 by using a power amplifier (PA) in the downlink direction (a higher transmission power could also be used at the base station) and by using a minimum value for the base station sensitivity level.

3.2.2.1 Downlink boosters and power amplifiers

In the early days of the GSM the radio link budget was typically limited by the uplink direction because the base station sensitivities were quite poor, compared to current products, and low noise amplifiers had high noise figures. Nowadays, new base station, diversity antenna and low noise amplifier products have shifted the radio link imbalance to the downlink direction. Thus, operators and equipment manufacturers have also started to focus on improving the downlink transmission power which can be done by increasing the base station transmission power or by reducing all the losses prior to the transmission antenna. In this context the first configuration is called an actual "booster" and it is co-located with the base station (the booster is typically implemented before the combiner) as shown in Figure 3.6. Moreover, the second configuration is called a "power amplifier" (PA) because this type of amplifier is installed next to the transmission antenna (see Figure 3.6). In the booster configuration the maximum output power is limited by the size of the booster. Base station equipment cannot be too large or take up too much space, due to site rental and installation costs and difficulties. Thus, the maximum power following the booster is a practical consideration and if the cable and other losses are not too high this booster configuration is satisfactory. However, the power amplifier gives flexibility in the downlink direction because even very high cable losses can be reduced in the downlink direction. Figure 3.6 shows that the power amplifier improves the downlink direction if the cable loss is larger than 3 dB (45–42 dBm, Figure 3.6) even if the booster is applied. It has to be noted that the utilisation of the power amplifier depends strongly on the base station configuration and on the cable length. If the power amplifier is used the booster is not needed.

UPLINK

Mobile station	Unit	Value	
RF power	dBm	33	A
Cable loss	dB	0	B
TX antenna gain	dBi	0	C
Peak EIRP	dBm	33	D = A – B + C
Base station	**Unit**	**Value**	
RX antenna gain	dBi	16	E
Diversity reception	dB	5	F
Depolarization loss	dB	0	G
Low noise amplifier	dB	5	H
Cable loss	dB	3	I
BTS sensitivity	dBm	– 108	J
Interference degradation margin	dB	1	K
Minimum reception level	dBm	– 130	L = – E – F + G – H + I + J + K
Isotropic path loss	dB	163	M = D – L

DOWNLINK

Base station	Unit	Value	
RF power	dBm	45.0	A
Combiner loss	dB	3	B
Power amplifier	dB	4	C
Cable loss	dB	3.0	D
TX antenna gain	dBi	16.0	E
Peak EIRP	dBm	59	F = A – B + C – D + E
Mobile station	**Unit**	**Value**	
RX antenna gain	dBi	0.0	G
Cable loss	dB	0.0	H
MS sensitivity	dBm	– 105.0	I
Interference degradation margin	dB	1.0	J
Minimum reception level	dBm	– 104.0	K = – G +H + I + J
Isotropic path loss	dB	163.0	L = F – K

Figure 3.5. The maximum power budget of the GSM900 system.

In order to realise the actual need for both these booster types a critical analyse of the power budget is needed, by taking into account the environment and establishing the configurations where boosters can improve the path loss. Figure 3.5 presented the power budget where the uplink direction is very good but also realistic and valid. This power budget shows that the booster is required in order to balance the path loss. It also has to be noted that only 3 dB cable loss is used in this calculation that means an additional 3 dB improvement by a power amplifier can be achieved in the downlink direction if the cable loss increases from 3 dB to 6 dB.

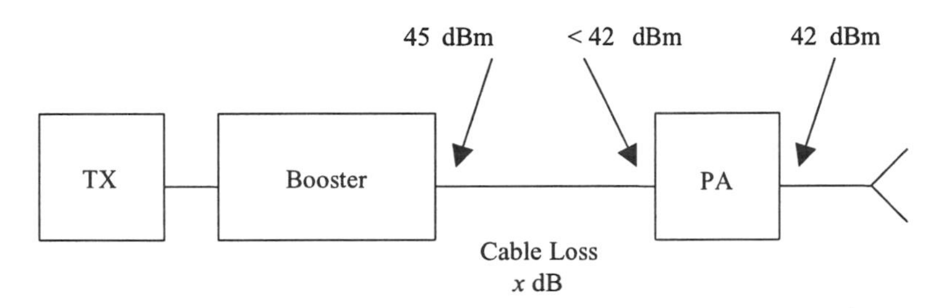

Figure 3.6. The downlink "booster" and power amplifier located in the antenna line set-up.

3.2.2.2 Frequency hopping and power budget

Frequency hopping is a function that enables the use of a pool of frequencies which are used in turn to decrease the co-channel interference and to improve the carrier-to-noise (*C/N*) or carrier-to-interference (*C/I*) ratio. Frequency hopping is thus used to decrease interference that corresponds to an increase of frequency reuse and that is called *interference diversity* (a capacity related function, see Chapter 6.3.2.2). Frequency hopping is also used to improve the received signal level against fading in a radio propagation channel, called *frequency diversity*.

At least two different frequency channels are required to be assigned to one base station in order to apply frequency diversity. The frequency diversity related improvement at the base station receiving end is called frequency diversity gain and it depends on the separation between two different frequency channels. This separation is called a coherence bandwidth and occurs when two different frequency channels provide uncorrelated signals (recall Chapter 1). The required coherence bandwidth is related to the environment: frequencies can be closer to each other in macrocellular environments than in microcellular or indoor environments to provide equal low correlation. The separation of 5–10 MHz,[10] is required in indoor locations to obtain signals of 0.7 correlation. These results are based on the measured values and tell that frequency diversity is available but the applicability is very low in indoor and microcellular locations because such a high frequency band is required to achieve the gain. Mobile operators have, typically, approximately 10 MHz frequency band at their disposal and thus this frequency diversity would be very difficult to realise. However, other studies have also been done and it is reported, for example,[11] that a reasonable improvement in an urban macrocellular environment can be achieved when the adjacent channels are already used. These results are based on GSM channel simulation application. These simulated results are, of course, only for the theoretical observation because the adjacent channels cannot be used in the same cells in the GSM system.

A final example of how frequency diversity is quite different compared to the other diversity methods; speed. After ≥ 3 km/h the frequency diversity disappears because the hopping rate in the GSM is not fast enough to achieve some diversity gain.[12] Moreover, when the mobile station is not moving, frequency hopping makes the radio propagation channel appear to move and it could be said that in the case of frequency hopping the *mobile station is always moving—either truly or virtually.*

3.2.2.3 Single and multi band antenna line configurations

An understanding of the radio propagation and power budget is the key to having a successful, cost-efficient and high quality single or multi band base station antenna line configuration. The configuration work starts from the analysis of the radio propagation which has a direct influence on the base station coverage area. It was previously mentioned and will later be explained in detail (in Chapter 4), that coverage areas are totally different in macro base stations at 450 MHz, 900 MHz, 1800 MHz and 2100 MHz frequency bands. Thus, there are two possible aims:

- *to have equal coverage areas for all the frequency bands*
- *to have a continuous coverage area for one frequency band with only hot spot coverage areas for the other frequency bands* (see Figure 3.7).

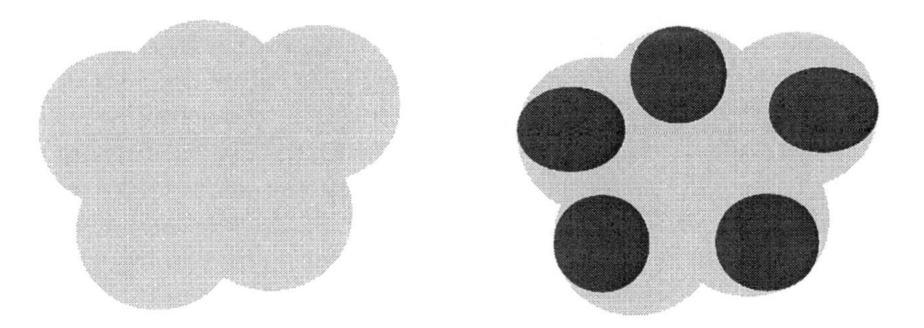

Figure 3.7. Coverage area targets for different frequency bands in the multi band radio network.

In the continuous coverage case, first, the base station sites for these different frequency bands *should not necessarily be co-located* in order to optimise coverage. Second, the base station antennas *should be installed at different heights* if the equal coverage area is required for all frequency bands and if these different frequency bands are used at the same base station site and they all have almost equal path loss in the power budget (as is typical in GSM). The path losses for the different frequency bands can also be modified so that the coverage areas are equal when using the same antenna height for all frequencies. However, the coverage area defines whether the same antenna height and location can be utilised for the different frequency bands. If the coverage areas can be different for the different frequency bands then co-located multi band antennas can be considered.

When the coverage areas for the different frequency bands are agreed, the power calculations have to be checked next in order to decide the antenna line configuration. The single band antenna line is always similar and the power budget calculations only define which equipment is implemented. The maximally equipped single band antenna line for the downlink and uplink directions is finally presented in Figure 3.8.

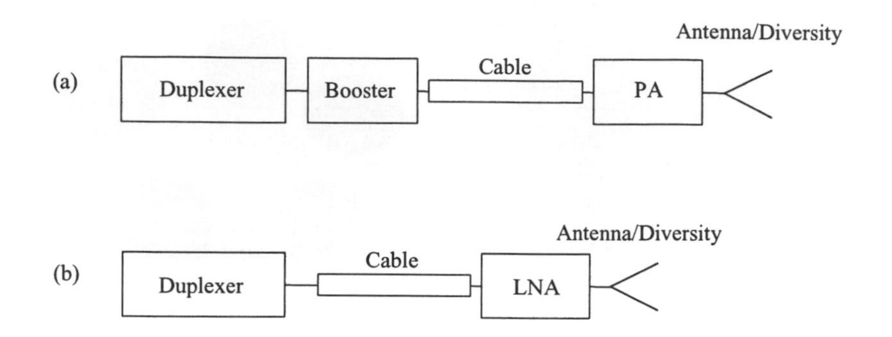

Figure 3.8. The fully equipped single band antenna line for the (a) downlink, and (b) uplink directions.

In the multi band configurations there are more alternatives to select and thus more detailed configuration planning is required. Equipment like diplexers, low noise amplifiers and diversity reception defines the integration of these different elements. Figure 3.9 shows equipment typically used in the downlink and uplink directions. It can be seen that diplexers are installed after the low noise amplifiers, excluding the usage of integrated dual band antennas and diplexers. Correspondingly, if the low noise amplifiers are not required at 900/1800 MHz the diplexer could be installed next to the antenna. Moreover, if the low noise amplifier is needed only at one frequency band (900 MHz or 1800 MHz) the configuration does not change because the frequencies cannot be combined before the amplifier. This analysis recommends the use of integrated elements containing multi band and dual polarised antennas and low noise amplifiers as well as diplexers, if possible. Thus, the main concern in the antenna line is to understand that the diplexer location depends on the low noise amplifier. When power amplifier is needed in the downlink direction the configuration will be more complex.

3.2.2.4 Duplex and diplex filter losses

Duplex filters are used just before the base station antenna connector to combine the main transmitting (TX) and receiving (RX) branches in order to connect the base station and the base station antenna by using only one cable. Thus, one cable, one antenna and the installation can be saved. Duplexers typically cause less than 1.0 dB additional loss together with the connectors in the uplink and downlink directions.

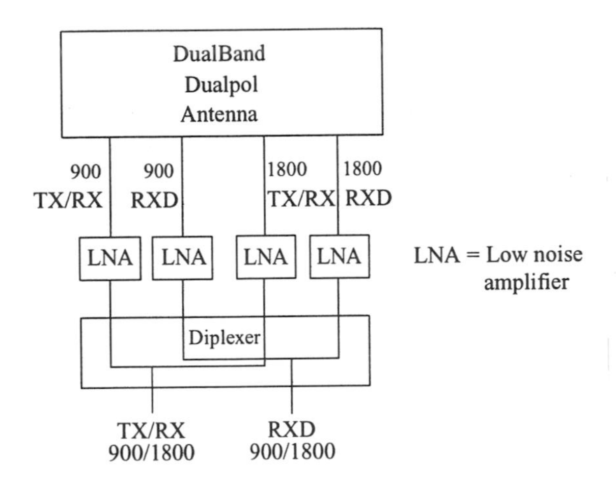

Figure 3.9. Dual band antenna line configuration with LNA.

Diplex filters are used just before the base station antenna connector and base station antenna (see Figure 3.9) to combine two transmitting (TX), receiving (RX) or combined TX/RX branches at different frequency bands in order to connect the base station and the base station antenna by using only one cable for each of them. Thus, one cable, one antenna, one diplexer and the installation is required for the dual/multi band TX, RX or TX/RX. Diplexers also typically cause less than 1.0 dB additional loss together with the connectors in the uplink and downlink directions.

3.2.2.5 Power budget margins

The power budget in Figure 3.5 contains one margin called *interference degradation margin* which is not the only margin in the radio coverage planning. In the power budget calculations the maximum allowed path loss is determined based on the base station transmitting peak power that is one parameter for the actual coverage prediction. The coverage prediction is made by using the software tools and the coverage area can be solved when the coverage threshold is determined. The coverage threshold can be determined by starting from the MS sensitivity (when the coverage prediction is done in the downlink direction) and by adding different margins to this value

Coverage threshold downlink = MS sensitivity + margins.

Thus, the final allowed path loss between base station and mobile station is much lower when all other margins are taken into account.

The value of 3 dB for the uplink and downlink for this interference degradation margin is recommended.[13] However, the explanation as to why this value is used is more or less inadequate. After several discussions with very experienced GSM engineers there can be several different and inaccurate explanations found for this margin. Two of these explanations are presented for consideration, though they may not be the right ones they are logical and intelligible.

It has been explained that the interference degrade margin is required because of the multiuser interference situations. This means that there are two different signals which cause interference (I) to the serving carrier (C) at a certain location, see Figure 3.10. When calculating the total C/I value the interference components I_1 and I_2 have to be added together in order to get the cumulative interference level $I = I_1 + I_2$. If the interference components I_1 and I_2 are received at the same phase and their received levels are equal the total interference level $I = I_1 + I_2$ is increased 3 dB. Thus, the total interference level (I) can be written $I = I_1 + 3$ dB $= I_2 + 3$ dB (when $I_1 = I_2$) and the interference degradation margin value 3 dB is defined based on these two interference case where interfering signals have the same phase and equal level at reception.

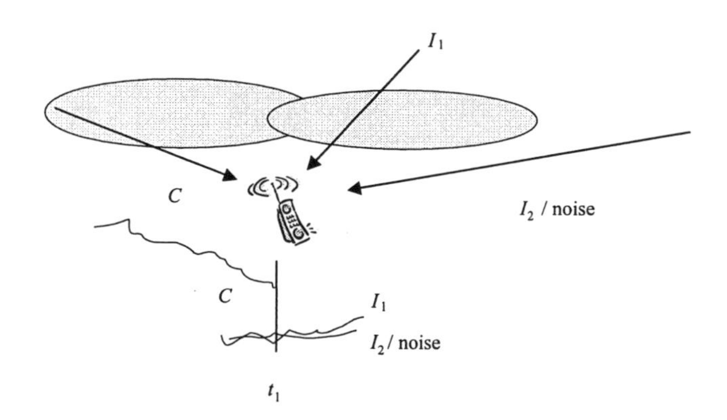

Figure 3.10. Two explanations of the interference degradation margin.

The second explanation is almost the same as the first but only one interference exists and it is counted with noise (the interference I_2 in Figure 3.10 is replaced by the noise).

As a conclusion it can be noted that this margin is not needed in the interference free areas (rural and urban areas). If there is still a willingness to use this margin it could at least be reduced down to 1 dB (as in the power budget in Figure 3.5) because the 3 dB situation is very theoretical.

3.3 Coverage and capacity optimized configurations

Having introduced the radio parameters that have a significant effect on the path loss between the base station and mobile station, it is still to be concluded which of these parameters are really required in the coverage and capacity driven configurations in order to manage calculation of the correct path loss for the coverage or capacity base stations. Table 3.8 shows the coverage maximized (only 1 frequency) and capacity maximized (≥ 2 frequencies) base station site configurations.

Table 3.8. Base station site equipment for the coverage and capacity needs.

Equipment	Coverage	Capacity
Combiner	By-passed	Wideband
Frequency hopping	No / Yes	Yes
TX peak power (W)	50	30
Cable	1 5/8"	1/2"
LNA	Yes	No
Diversity	Yes	No
Booster	Yes	No
Dual band antennas	No	Yes

It can be noted that the base station configurations are quite different for the coverage and capacity needs. Hence, good configuration planning must be made for each area and for each purpose.

In this configuration planning the *network evolution* also has to be remembered because the power budget can be maximized, for example, by using power amplifiers. Recall that power amplifiers are frequency selective (only one frequency can be transmitted) and thus the capacity cannot be easily increased. However, the base station can be extended from the one TRX configuration to two TRX by using polarisation diversity antenna and transmitting the first TRX by using the first polarisation and transmitting the second TRX by using the second polarisation (different transmission polarisations represent different actual antenna lines). Power amplifiers can also be utilised (one per TRX) and the coverage area can be maximized. This is *evolution planning* that is really needed for long term cost efficient configurations.

3.4 Conclusions

Configuration planning and power budget has a great effect on coverage planning and many links to capacity and frequency planning. These links between the configuration planning and capacity/frequency planning influence the total radio network configuration. Capacity related issues like capacity functionality (frequency hopping) or the base station configuration (number of TRXs versus need of combiner) have a direct influence on the base station site configuration and thus on the base station coverage radius. Table 3.9 reviews configuration planning for maximising base station coverage area at a minimum cost.

Table 3.9. Configuration planning and power budget.

Subject	**Findings**
Power budget calculations	–Maximize the allowed path loss –Balance the power budget by tuning the BTS TX RF peak power
BTS and MS transmission power	–Use supplier related nominal values
BTS and MS sensitivities	–Use supplier related nominal values
Combiner losses	–Use supplier related nominal values
Base station antennas	–Use mostly narrow vertical beamwidth
Base station antenna installations	–Remember backlobe and antenna height
Cables and connectors	–Frequency band selective
Low noise amplifiers	–NF = 1.5 dB, Gain = 14 dB, cable loss < 5 dB –NF = 1.5 dB, Gain = 21 dB, cable loss > 5 dB
Diversity reception	–Polarization diversity in the urban areas –Space diversity in the rural areas
Downlink boosters	–Use PA to reduce the downlink cable loss
Frequency hopping	–Frequency diversity helps when mobile is not moving
Single band antenna line Multi band antenna line	–LNA for both main and diversity RX branch –Define coverage areas for the different bands –Define power budgets for the different bands –Plan the multi band antenna line carefully –Dual band antenna with diplexers ≠ LNA
Duplex filter losses	–Use supplier related nominal values
Power budget margins	–Interference degrade margin can be diminished

3.5 References

[1] ETSI, Digital cellular telecommunications system (Phase 2+), Radio transmission and reception, GSM 05.05.

[2] Warren L. Stutzman, Gary A. Thiele, "Antenna Theory and Design," John Wiley & Sons, 1998.

[3] Kazimierz Siwiak, "Radiowave Propagation and Antennas for Personal Communications," Artech House, 1998.

[4] Jyri S. Lamminmäki, Jukka J.A. Lempiäinen, "Radio Propagation Characteristics in Curved Tunnels" IEEE—Microwaves, Antennas and Propagation, August 1998, vol. 145, no. 4, pp. 327–331.

[5] D.A. Bell, "Electrical Noise," D. Van Nostrand Company, 1960, London.

[6] W. Lee, "Mobile Cellular Telecommunication Systems," McGraw-Hill Book Company, 1990, p. 449.

[7] D. Parsons, "The Mobile Radio Propagation Channel," Pentech Press, 1992.

[8] J.J.A. Lempiäinen, "Assessment of diversity techniques in a microcellular radio propagation channel," Doctoral Dissertation, Helsinki University of Technology, 1999.

[9] W. Lee, "Mobile Communications Design Fundamentals," John Wiley & Sons, 1993, p. 372.

[10] J-F. Lemieux, M.S. El-Tanany, H.M. Hafez, "Experimental Evaluation of Space / Frequency / Polarisation Diversity in The Indoor Wireless Channel," IEEE Transactions on Vehicular Technology, August 1991, vol. 40, no. 3, pp. 569–574.

[11] J. Haataja, "Effect of Frequency Hopping on the Radio Link Quality of DCS1800/1900 System," Masters Thesis, Helsinki University of Technology, 1997.

[12] A. Nieminen, "Influence of Frequency Hopping on the Frequency Planning in Indoor Locations", BSc Thesis, Häme Polytechnics, 2000.

[13] ETSI, Digital cellular telecommunications system (Phase 2+), Radio network planning aspects, GSM 03.30.

[14] W.C. Jakes, Jr., (ed.), "Microwave Mobile Communications," Wiley-Interscience, 1974.

Chapter 4

COVERAGE PLANNING CRITERIA

4. COVERAGE PLANNING CRITERIA

4.1 Location probabilities and fading margins

The characteristics of a radio path, the environment, the behaviour of the users, the weather, etc. cause uncertainty when receiving radio signal. In commercial radio networks the aim is not usually to achieve the maximum performance but to guarantee a certain quality of service with high probability. The objective in commercial radio networks is to optimise performance from the end-user's point of view and the revenue of the network. This means that time and location probabilities for the network must be high enough to ensure customer satisfaction.

Quality of radio coverage in a mobile network is usually measured as location probability and time probability. Location probability defines the probability in which the field strength is above a sensitivity level in the target area. A location probability of 50 percent corresponds to the situation in which the average field strength equals the sensitivity of a receiver over a certain geographical area. Of course, the desired probability is usually higher than 50 percent, that is, the field strength should be higher than the sensitivity of a receiver. Time probability is defined over a longer period to account for phenomena in which field strength is not constant. For example, leaves in trees can lower field strength at street level during summer time and when winter comes the average field strength increases. Also, weather (for example rain) can cause a fluctuation in field strength and these changes can be quite remarkable. Because time probability is even more difficult to measure and predict than location probability it is usually taken into account by choosing tighter thresholds; sometimes it is not considered at all. However, location and time probabilities are among of the most important variables in coverage planning.

4.1.1 Slow (log-normal) fading and standard deviation

When trying to build good coverage in built-up areas the challenge is the fact that mobile antennas are usually well below the height of surrounding buildings, and there is no line of sight path between the base station and the mobile station. This means that propagation happens mainly by reflections from the surfaces of the buildings and by diffraction over and around them. Figure 4.1 illustrates some possible mechanisms of propagation in an urban area.

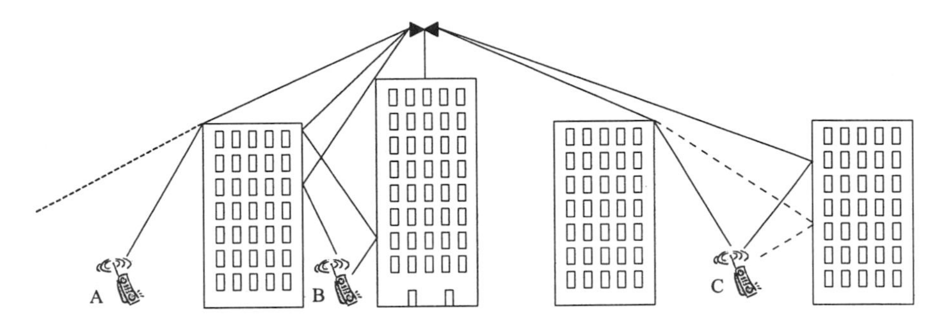

Figure 4.1 Components of fading. Slow fading is caused by the obstacles near the mobile stations. All three mobiles (A, B, C) are suffering slow fading because they are in the shadows of the buildings.

Slow fading, often called shadowing, is caused by the mobile moving into the shadow of hills, trees or buildings. There is no satisfactory model for slow fading, but according to various measurements the mean path loss closely follows a log-normal distribution. Therefore slow fading is sometimes called log-normal fading.[1]

The slow fading describes the variations of the average signal strength due to obstacles on the signal path. Of course, the term "average signal strength" is used here with a special meaning. The correct term would be "sliding average of signal." The fast fading component in a signal can be extracted by using an average window of thirty-two wave length.[1] If using the sliding average window, the result is a signal not having a fast fading component but including the shadowing effect. It is reported that a signal where fast fading is filtered follows log-normal distribution. If the slow fading signal is expressed in decibels the distribution of the signal follows normal distribution. Therefore, the slow fading signal can be presented with a normal distribution as shown in the following equation.

$$p_x = \frac{1}{\sqrt{2\pi\sigma^2}} e^{-\frac{(x-\mu)^2}{2\cdot\sigma^2}}$$ *Equation 4.1*

In Equation 4.1 x is the random variable (here the slow fading signal), μ is the mean value of x and σ is the standard deviation of x. The value of standard deviation σ depends on the environment close to the receiver.

The standard deviation for the log-normal fading in urban areas is stated to be around 8 dB.[1]

When measuring the signal strength without fast fading (averaged signal) at the given distance, the values are concentrated close to the mean value μ. Values follow log-normal distribution. Following the definition for standard deviation approximately 68 percent of samples fall between $\mu \pm \sigma$. The following example is a representation of the idea.

Example 1

The average field strength at the given distance is –82 dBm. Let us assume that σ is 8 dB.

Average (μ)	=	–82 dBm
Deviation (σ)	=	8 dB
$\mu - \sigma$	=	–82 dBm – 8 dB = –90 dBm
$\mu + \sigma$	=	–82 dBm + 8 dB = –74 dBm

Approximately 68 percent of samples fall between –90 dBm and –74 dBm.

Example 1 indicates that 68 percent of the signal strength samples fall in the range between $\pm\,\sigma$ around the mean value μ. Of course, if field strength is measured in a certain area μ and σ can be determined.

4.1.2 Slow fading margin

When the distribution of field strength is defined the next step is to determine the *slow fading margin*. When planning coverage for a mobile network, one quality target is location probability of coverage. This means that field strength must be better than the given threshold with some probability. As discussed earlier field strength fluctuates due to the shadowing effect—sometimes a signal level is above average but sometimes below the average. To guarantee a certain minimum signal level in a radio network we have to introduce a margin against slow fading. This margin is called a slow fading margin. We will continue Example 1 to determine a slow fading margin.

Example 2

In Example 1 the following situation was presented. The mean value of the signal is –82 dBm. If the standard deviation of the signal is 8 dB, approximately 68 percent of the samples fall between –90 dBm and –74 dBm.

If the sensitivity of a mobile phone is –100 dBm we can calculate the probability for the situation in which the field strength is below -100 dBm.

$$P(x \le -100\,\text{dBm}) = \int_{-\infty}^{x} \frac{1}{\sqrt{2\pi}\sigma} e^{-(1/2)[(t-\mu)/\sigma]^2} dt =$$

$$\int_{-\infty}^{-100} \frac{1}{\sqrt{2\pi}\cdot 8} e^{-(1/2)[(t+82)/8]^2} dt = 0.012224$$

As it can been seen in the equation above the probability for the situation in which the received power is below –100 dBm is approximately 1.2 percent or the location probability at the given distance is 98.8 percent (100 percent - 1.2 percent).

Slow fading margin can be seen as a way to improve probability of having a high enough signal level. If we accept that there is a 50 percent probability that the signal level is below the sensitivity level, that is, there is no need for a slow fading margin, then the average signal level can remain at –100 dBm. Of course, 50 percent probability in a commercial network is far too low. In Example 2 we have increased the probability from 50 percent to 99 percent by introducing a slow fading margin of 18 dB. Slow fading margin depends on the location probability as shown in Figure 4.2.

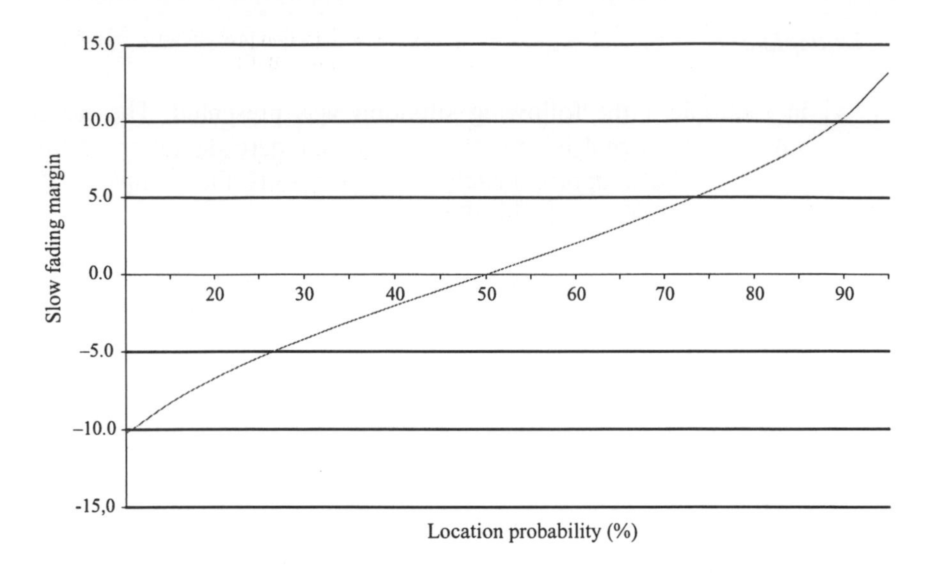

Figure 4.2. Slow fading margin as a function of location probability. σ = 8 dB.

In Figure 4.2, the slow fading margin:

- is negative if location probability is below 50 percent,
- grows towards infinity when location probability grows close to 100 percent,
- is 0 dB if location probability is 50 percent.

The previous calculation applies for the locations on the cell edge. In radio networks location probability and slow fading marginal are calculated over the entire cell. This changes the approach because the signal level usually becomes stronger closer to the transmitter, requiring clarification of the relationship between area location probability and point location probability.

4.1.3 Point location probability

Having shown that the slow fading margin is 0 dB when the *point location probability* is 50 percent. If better better location probability is required a higher slow fading margin is applied. This is described in Figure 4.3 where the smaller circle is decreased cell size due to a slow fading margin of 5.1 dB.

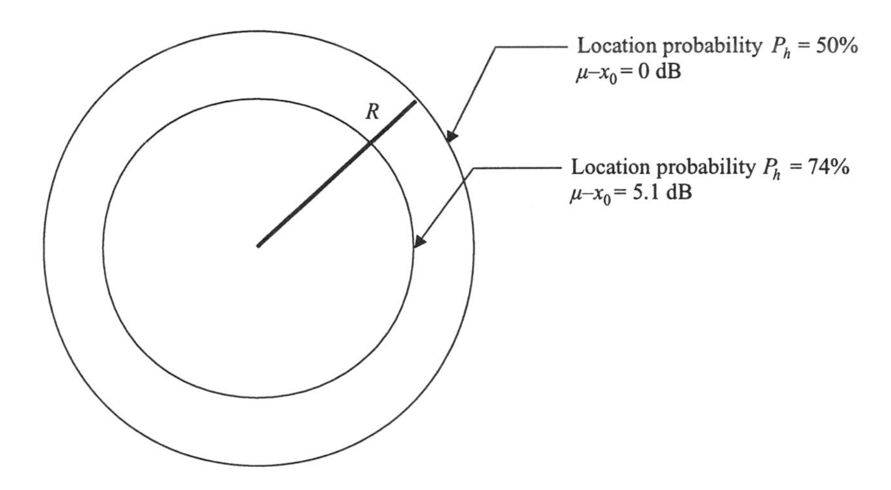

Figure 4.3. Description of slow fading margin. P_h is point location probability, x_0 the minimum accepted signal level (e.g. sensitivity level) and μ is the mean value of x_0.

If point location probability on cell edge, P_h, increases from 50 percent to 74 percent then slow fading margin increases from 0 dB to 5.1 dB (when σ is 8 dB) and therefore the cell range gets shorter.

4.1.3.1 Area location probability

Outdoor location probability is a parameter that describes the probability that the received signal strength exceeds the minimum needed signal strength at the receiver in order to make a successful phone call outdoors. Point location probability can be converted to area location probability which means that probability of coverage is calculated over the cell. A_{useful} defines the useful service area within a circle of radius R where the signal level received by the mobile station exceeds a given threshold x_0. If p_{x_0} is the probability that the received signal, r, exceeds x_0 in an area dA, then the useful area can be written,[2] as

$$A_{useful} = \frac{1}{\pi R^2} \int p_{x_0}\, dA \qquad \textit{Equation 4.2}$$

where

$$p_{x_0} = \int_{x_0}^{\infty} \frac{1}{\sqrt{2\pi\sigma^2}} e^{-\frac{(r-\mu)^2}{2\cdot\sigma^2}} d\bar{r} - \frac{1}{2} - \frac{1}{2} erf\left(\frac{x_0 - \mu}{\sigma \cdot \sqrt{2}}\right) \qquad \textit{Equation 4.3}$$

In a macrocellular environment the mean value of the received signal strength, μ, can be written as

$$\mu = F_{cell\ edge} - \gamma \log_{10}\left(\frac{d}{R}\right) \qquad \text{Equation 4.4}$$

where $F_{cell\ edge}$ is field strength on the cell edge in dBμV/m and d is the distance from base station to some point inside the cell coverage area, R is the distance from the base station to the cell edge and γ is propagation slope (recall Chapter 1). The value of γ normally varies from 25 dB/dec to 45 dB/dec. In propagation models γ is also used as a key parameter. For example, in Okumura–Hata formula the value of γ is expressed as shown in Equation 4.5. As can be seen γ depends on the height of the base station antenna h_{BTS}.[3]

$$\gamma = 44.9 - 6.55\log_{10}(h_{BTS}) \qquad \text{Equation 4.5}$$

Substituting μ by Equation 4.4 probability P_{x^0} can be presented as

$$p_{x_0} = \frac{1}{2} - \frac{1}{2}\cdot erf\left(\frac{x_0 - F_{cell\ edge} + 10\cdot\gamma\cdot\log_{10}\left(\frac{r}{R}\right)}{\sigma\cdot\sqrt{2}}\right) \qquad \text{Equation 4.6}$$

If making the following substitutions

$$a = \frac{(x_0 - F_{cell\ edge})}{\sigma\sqrt{2}}$$

$$b = 10\gamma\log_{10}\left(\frac{e}{\sigma\sqrt{2}}\right) \qquad \text{Equation 4.7}$$

Equation 4.2 can be written as

$$A_{useful} = \frac{1}{2}\left\{1 + erf(a) + e^{\left(\frac{2\cdot a\cdot b+1}{b^2}\right)}\left[1 - erf\left(\frac{a\cdot b+1}{b}\right)\right]\right\} \qquad \text{Equation 4.8}$$

Using Equation 4.8 the location probability over whole cell coverage area can be calculated. The difference between location probability on the cell edge (point location probability) and area location probability can be seen in Figure 4.4.

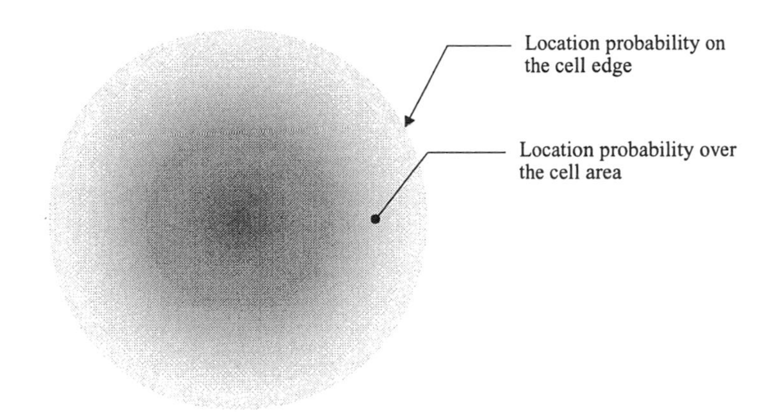

Figure 4.4. Point location probability is defined in a point on the cell edge. Location probability over a cell area takes into account the fact that the signal level gets higher closer to the transmitter.

Slow fading margins for stardard deviations of 6 dB, 7 dB and 8 dB are presented in Table 4.1.

Table 4.1 Slow fading margins. The margins are calculated for point location probability and area location probability with standard deviations of 6 dB, 7 dB and 8 dB. In case of area location probability slope is assumed to be 33.8 dB/dec.

Location Probability (%)	**Standard deviation (dB)**					
	Point location probability			**Area location probability**		
	6	**7**	**8**	**6**	**7**	**8**
10	−7.69	−8.97	−10.25	−19.34	−20.19	−21.12
20	−5.05	−5.89	−6.73	−14.14	−14.85	−15.59
30	−3.15	−3.67	−4.20	−10.92	−11.46	11.99
40	−1.52	−1.77	−2.03	−8.43	−8.77	−9.09
50	0.00	0.00	0.00	−6.25	−6.39	−6.51
60	1.52	1.77	2.03	−4.18	−4.11	−4.01
70	3.15	3.67	4.20	−2.08	−1.76	−1.41
80	5.05	5.89	6.73	0.28	0.90	1.55
90	7.69	8.97	10.25	3.43	4.48	5.56
95	9.87	11.51	13.16	5.95	7.36	8.80
96	10.50	12.25	14.01	6.67	8.18	9.73
97	11.28	13.17	15.05	7.55	9.19	10.87
98	12.32	14.38	16.43	8.72	10.53	12.38
99	13.96	16.28	18.61	10.54	12.62	14.74

Indoor location probability

Indoor coverage location probability describes the probability that the received signal strength exceeds the minimum needed signal strength at the receiver inside buildings. The principle of calculation is very similar to outdoor. The differences between these two location probability calculations are

- when calculating indoor location probability standard deviation is replaced by standard deviation of signal measured indoors,
- average building penetration loss is added to the calculated slow fading margin.

Because standard deviation of signal indoors is higher than standard deviation outdoors the slow fading margin is higher. Indoor building penetration loss should also be added which results in much higher slow fading margins.

4.1.4 Multiple server location probability

A radio network consists of several base stations. In some locations there may be more than one server that can provide good coverage. For the previous calculations the key assumption was a *single, isolated cell.* However, if there are several cells providing coverage in an area the probability for having high enough field strength increases. The situation is illustrated in Figure 4.5.

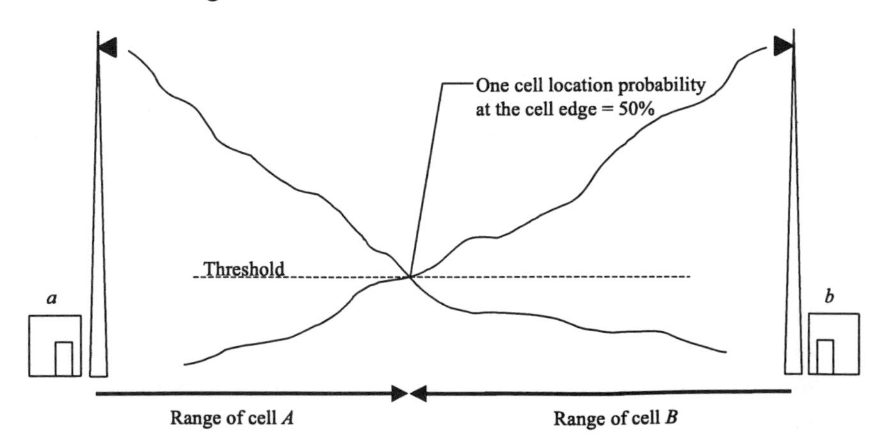

Figure 4.5. Received signal strengths for cell a and cell b. The dotted line presents the coverage threshold.

Assuming there are two cells providing coverage in the target area, at the cell edge both cells are providing 50 percent location probability (50 percent probability that field strength is above or below the given threshold). Furthermore, if assuming that the signal from the cells are uncorrelated a joint probability could be calculated as:

$$(A+B)-(A\cdot B) \qquad \textit{Equation 4.9}$$

In Equation 4.9 *A* is point location probability of cell *a* and *B* is the point location probabilities of cell *b*. For example if point location probability is 50 percent for both cells (slow fading margin is 0 dB) the two-server coverage location probability at the cell edges is

$$(A+B)-(A\cdot B)=(50\%+50\%)-(50\%\times 50\%)=75\% \qquad \textit{Equation 4.10}$$

As Equation 4.10 shows the location probability increases from 50 percent to 75 percent if there are two servers having an equal signal level at the point under study and the signals from the cells are uncorrelated.

The method presented shows the principle of multiple server location probability. Calculation of location probability over the network is quite complicated. First, propagation should be known, as the slope affects the location probability. Second, network layout partly determines how smooth coverage is. Finally, real radio environments make calculation almost impossible. Although, it is nearly impossible to calculate multiple server location probability in a live network it can be calculated for theoretical cases. In Figure 4.6 point location probability is calculated as a function of single cell location probability for two-server and three-server cases. Note that it is assumed that correlations between the signals from the neighbours are zero. This assumption may be difficult to fulfill in live radio networks.

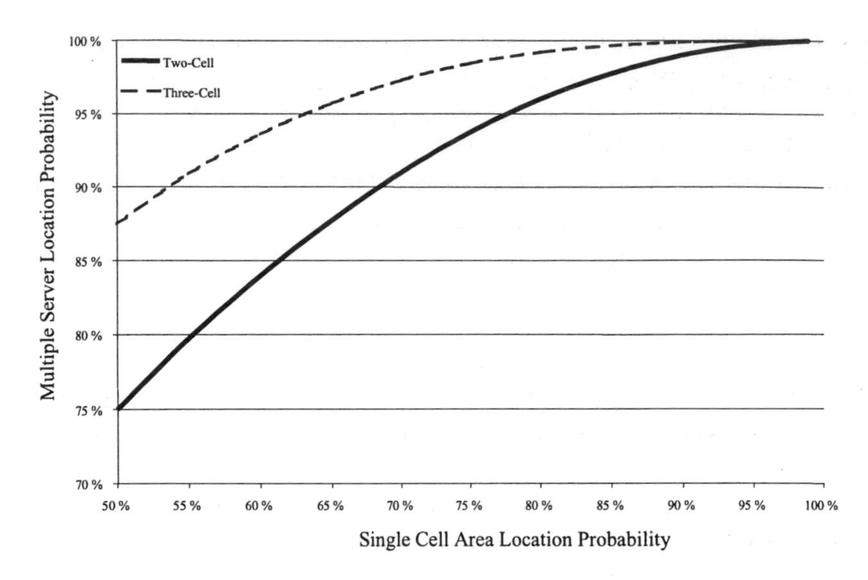

Figure 4.6. Results of multiple server coverage location probability.

As can be seen from Figure 4.6 the multiple server coverage location probability approaches the value of 100 percent logarithmically as the single cell area location probability increases. In GSM the full benefit of multiple servers cannot be utilised due to slow handover procedures through some gain can be expected. In CDMA soft handover enables utilisation of the full gain.

4.2 Building and vehicle penetration

Building penetration loss is loss due to roofs, walls and floors. Building penetration loss varies according to building type. When indoor coverage is important, the average building penetration loss and standard deviation indoors should be known to calculate corresponding slow fading margins. If measurement results are not available the values can be estimated, for example, according to the ETSI GSM recommendation 03.30 in which building penetration loss in urban areas is estimated to be 15–18 dB and in rural areas 10 dB depending on the frequency used.[4] For standard deviation indoors default values of 9–11 dB can be used. It should be noted that some studies have shown that the average penetration loss for some types of building decrease when the frequency is increased, e.g. from 900 MHz to 1800 MHz.

Building penetration loss varies due to the effect of the following conditions:

- Different types of outside wall construction
- Changes in floor elevation
- Propagation environment near the building
- Different percentage of window areas in the outside walls
- Different building orientations with respect to the direction of antenna illumination
- Different types of window treatment used to reflect sunlight and heat.

Building penetration loss can be verified with a measurement campaign. However, to get statistically reliable results the number of measured building should be remarkable. Of course buildings can be classified according to size, material, usage, etc. and measurements can be concentrated to the most predominant ones.

To avoid fast fading the measurement locations should be planned so that at least thirty-two samples can be collected within 32 λ. A local average with thirty-two samples is used to eliminate the fast fading effect due to multipath propagation. To obtain a reference signal level for building penetration loss, measurements along the outside perimeter of the buildings should be made at street level (MS antenna height 1.5 m). Building penetration loss for a given floor area is defined to be the decibel difference between the average of the indoor measurements and the average of measurements measured at street level (see Table 4.2).

Table 4.2 Building penetration losses (BPL) in different buildings (example).

	Floor										
Building 1	Basement	1st	2nd	3rd	4th	5th	6th	7th	8th	9th	10th
BPL (dB)	40	31	42	39	37	27	24	15	—	—	—
Building 2	Basement	1st	2nd	3rd	4th	5th	6th	7th	8th	9th	10th
BPL (dB)	22	—	22	—	20	—	20	—	19	—	17

In practical radio planning, building penetration loss is an important parameter in predicting radio coverage. This parameter is used to determine the coverage threshold and the indoor coverage cell range with a certain probability. If the aim is to have very good coverage in lower floors the planning threshold must be higher than when compared to cases where the important floors are higher.

Vehicle penetration is similar to building penetration in radio network planning. The margin *car penetration loss* is needed to compensate the loss due to the bodywork of a car. The loss depends on the materials used in its manufacture. The smallest losses are achieved if the body is made of plastic. Car penetration losses vary typically between 5 to 15 dB.

4.3 Other losses and margins

In addition to the slow fading margin and building penetration losses there are several margins that are used to ensure proper coverage. The usage of these margins must be considered case by case;

- body loss
- antenna orientation loss
- polarisation loss
- additional fast fading margin
- interference degradation margin.

Body loss or body proximity loss affects both uplink and downlink. A human body near the mobile antenna affects the radiation pattern of the MS antenna. The loss measured against the best direction depends on the MS antenna type, location of the antenna and frequency. For planning purposes the loss of 4–6 dB could be used. However, in the worst direction the loss can be well over 10 dB.

In GSM sensitivity is measured in static, TU50, HT100 and RA250 fading profiles (900 MHz).[5] If planning cells for other environments it is possible to optimise sensitivity figures. Usually only done for a microcellular environment, where the speed of a mobile is slow and propagation suffers from fast fading. This can be taken into account by introducing an additional fast fading margin to correct the sensitivity figure determined from GSM recommendations.

An interference degradation margin is mentioned in GSM recommendations, and is used in cases where interference starts to increase above background noise and thus reduces the coverage area of cells. Sometimes this margin is also used for man-made interference.

4.4 Radio planning parameters in a network planning system

Radio link power budget deals mainly with the performance of radio hardware but includes some aspects of environment. Common environment related parameters are sensitivity and diversity gain (recall Chapter 3).

4.4.1 Power budget parameters

When planning coverage all of the details in a radio link power budget can be covered by two factors

- equivalent isotropic radiation power (EIRP)
- radiation pattern of base station antenna.

All common propagation models work in the downlink direction which means that a base station has a certain EIRP and transmitting antenna. If uplink is much weaker than downlink this method may lead to a situation in which downlink is fulfilling the planning criteria but uplink is suffering bad quality. To avoid this the downlink direction must be adjusted according to uplink performance. This is called *balancing*.

When entering data into a planning system only the downlink values are normally used. There are two important tasks in entering values. First, the antenna must be selected correctly. If the wrong radiation pattern is used, coverage prediction will give the wrong results and this causes problems in capacity planning, frequency planning, etc. After selecting the antenna the EIRP of the base station has to be set. This can be done by setting TX power, gains and losses in a way that the EIRP is achieved.

4.4.2 Thresholds

Coverage thresholds in network planning systems are usually presented with different colours. These thresholds are calculated based on slow fading margins, building penetration losses and vehicle penetration losses. Table 4.3 shows how planning thresholds are defined for different coverage targets. In indoor cases both building penetration loss (15 dB) and slow fading margin are added to sensitivity (–100 dBm in this example). The difference between indoor and outdoor thresholds is 18.3 dB if the target is 90 percent location probability. The difference is greater than building penetration loss 15 dB because standard deviation in the indoor case is higher (10 dB > 7 dB). This causes the increase in the slow fading margin.

Table 4.3. Coverage thresholds in planning system.

	Indoor			Outdoor (BPL = 0 dB)		
Location probability over cell area (%)	90.0	80.0	70.0	90.0	80.0	70.0
Slow fading margin + BPL (dB)	22.8	17.9	14.4	4.5	0.9	–1.8
Coverage threshold (dBm)	–77.2	–82.1	–85.6	–95.5	–99.1	–101.8

When plotting coverage in planning systems colours or grey scale shading can be used to present different coverage areas (thresholds from Table 4.3.) as shown in the Figure 4.7.

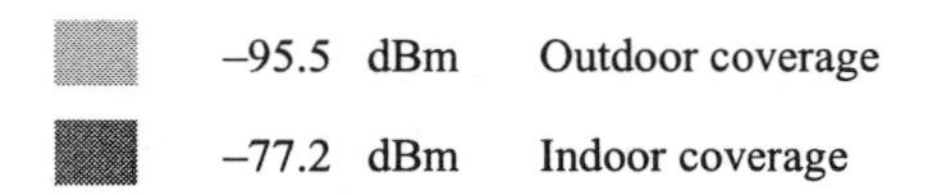

Figure 4.7. Coverage thresholds in a planning system.

4.5 Conclusions

This chapter examined coverage planning criteria. First, location probability and slow fading margins were discussed. Usually coverage probability is expressed with location probability; a way to improve location probability is to introduce a margin that reduces cell range but at the same time improves the average field strength over the cell area. This margin is called "slow fading" because deviation of field strength is "slow." Second, indoor and in-car coverage were discussed and building penetration loss was introduced. Finally, the coverage planning thresholds were determined. Table 4.4 gathers the topics of the coverage planning criteria.

Table 4.4. Coverage planning criteria.

Subject	**Findings**
Location probability	–Defines probability when signal is over a given threshold over a cell area –There can be different definitions
Slow fading	–Due to obstacles like buildings, hills, etc. –Also called log-normal fading
Slow fading margin	–Used to increase location probability –Standard deviation affect the margin
Building penetration loss	–Varies due to construction materials, thickness of walls, etc.
Coverage thresholds	–The targeted field strength value in planning system
Link budget balancing	–Coverage planning must be done according to the weaker direction
Downlink prediction	–Almost all propagation models work in the downlink direction

4.6 References

[1] W.C.Y. Lee, "Mobile Cellular Telecommunication Systems," McGraw -Hill Book Company, 1990, 449 p.

[2] W.C. Jakes, Jr., (ed.), "Microwave Mobile Communications," Wiley-Interscience, 1974.

[3] Hata, M., "Empirical Formula for Propagation Loss in Land Mobile Radio Services," IEEE Transactions on Vehicular Technology, vol. VT-29, no. 3, August 1980, pp. 317–325.

[4] ETSI, Digital cellular telecommunications system (Phase 2+), Radio network planning aspects, GSM 03.30.

[5] ETSI, Digital cellular telecommunications system (Phase 2+), Radio subsystem link control, GSM 05.05.

Chapter 5

RADIO PROPAGATION PREDICTION

5. RADIO PROPAGATION PREDICTION

Radio propagation prediction is one of the fundaments for radio network planning. If network planning is carried out with the help of a network planning system then coverage planning, frequency planning, capacity planning, interference analysis, dominance analysis, handover analysis, etc. rely on the propagation predictions. It is thus vital that radio propagation predictions are as accurate as possible taking into account the practical limitations. Sometimes radio propagation predictions are seen as a necessary phase in network planning but the importance of predictions and their influence in other phases are not well recognised.

5.1 Elementary propagation fundamentals

There are three mutually independent propagation phenomena. Fast fading causes rapid fluctuations in phase and amplitude of a signal if a transmitter or receiver is moving or there are changes in the radio environment (e.g. car passing by). If a transmitter or receiver is moving, the fluctuations occur within a few wave lengths. Because of its short distance fast fading is considered as small-scale fading.

Slow fading occurs due to the geometry of the path profile. This leads to the situation in which the signal gradually gets weaker or stronger. For example, a building near the base station causes shadowing in a relatively small area. The diameter of the area can be tens or, to a maximum, several hundreds of meters. Because there is fast fading on top of slow fading, a slow faded signal is usually measured over 32 λ. This averages fast fading and gives a slow-faded local-mean value for the signal.[1]

Large-scale attenuation depends on the path profile of the radio link. Usually the loss is expressed as a function of antenna height, frequency and the distance between the transmitter and receiver. Static propagation models do not usually take into account the attenuation caused by terrain but they express the loss. Of course, propagation models can be more accurate if terrain data is used in prediction. The mechanisms in large-scale attenuation are free space loss, groundwave propagation and diffraction.

In free space, the energy of radio waves diverges over an area proportional to the square of the propagation distance. Thus the free space loss can be expressed as in Equation 5.1.

$$L_{free\ space} \sim 20\log_{10}(d) \qquad \text{Equation 5.1}$$

Equation 5.1 gives the loss in decibels. In Equation 5.1 d is distance to the transmitter.

5.1.1 Reflections, diffractions, absorptions

When radio waves travel over land they interact with the earth's surface. The received signal is a complex sum of all its components. Equation 5.2 presents the sum of the components.

$$E = E_0\left(1 + R_c e^{i\omega} + (1 - R_c)F(\bullet)e^{i\omega} + \ldots\right) \qquad \text{Equation 5.2}$$

In Equation 5.2 E is the received signal, E_0 is theoretical field strength for propagation in free space, R_c is the reflection coefficient, $F(\bullet)$ the complex surface wave attenuation, ω is the phase difference between the reflected and the direct wave and i is the imaginary number $\sqrt{-1}$. Equation 5.2 can be simplified if used in high frequencies because $|F(\bullet)| \ll 1$ is valid for most situations. This assumption can be made if using a typical antenna height in transmitting and receiving ends. The simplified equation is

$$E \approx E_0\left(1 + R_c e^{i\omega}\right) \qquad \text{Equation 5.3}$$

and path loss of the presented signal is

$$L \sim 20\log\left(\left|1 + R_c e^{i\omega}\right|\right) \qquad \text{Equation 5.4}$$

In most cases ($d >> h_{BTS}h_{MS}$, where h_{BTS} is height of transmitter antenna and h_{MS} is the height of receiving antenna), the phase difference ω can be approximated by Equation 5.5.

$$\omega \approx \frac{4\pi h_{BTS} h_{MS}}{d\lambda} \qquad \text{Equation 5.5}$$

Also, the reflection coefficient can be approximated by assuming that $R_c \rightarrow -1$. This assumption is valid for most terrain types. If studying propagation further from the transmitter, it can be assumed that $d\lambda >> h_{BTS}h_{MS}$. Then the total path loss for Equation 5.2 can be approximated by

$$L \approx 20\log_{10} \frac{h_{BTS} h_{MS}}{d^2} \qquad \text{Equation 5.6}$$

Equation 5.6 shows that field strength diminishes proportionately with the inverse of the fourth power of the distance. According to experiences in a real network the result is correct. Closer to the transmitter the direct signal dominates the path loss and thus the slope of the signal is smaller i.e. the decrease with the signal is less rapid.

If the direct line of sight between transmitter and receiver is obstructed, radio waves attenuate due to diffraction. Diffraction is studied and modelled by assuming a knife-edge obstacle is located perpendicular to the line of sight in free space (see Figure 5.1).[2]

When calculating diffraction loss a diffraction parameter v is needed.

$$v = h_m \sqrt{\frac{2}{\lambda}\left(\frac{1}{d_t} + \frac{1}{d_r}\right)} \qquad \text{Equation 5.7}$$

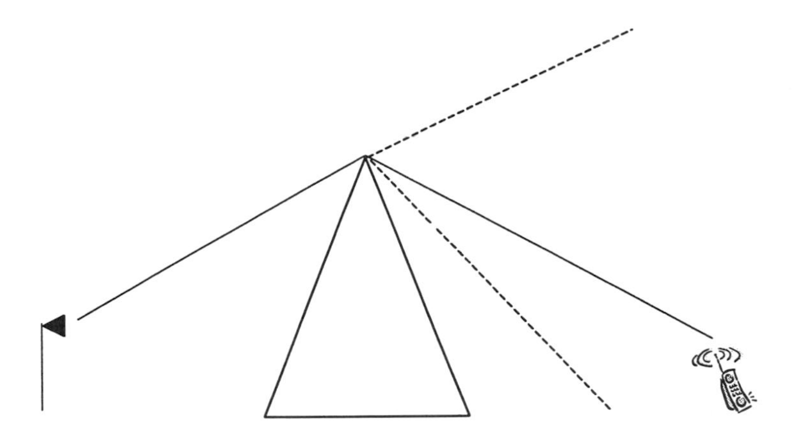

Figure 5.1 Diffraction in knife-edge obstacle.

In Equation 5.7 h_m is the height of the obstacle, and d_t and d_r are the terminal distances from the obstacle. The diffraction loss A_d can be approximated by

$$A_d = \begin{cases} 6 + 9v - 1.27v^2 & 0 < v < 2.4 \\ 13 + 20\log_{10} v & v > 2.4 \end{cases} \qquad \text{Equation 5.8}$$

The total radio path loss where a knife-edge obstacle exists is the sum of free space loss and diffraction loss i.e. $A_{total} = A_{free\ space} + A_d$. Single edge approximation is not very accurate in 900 MHz, 1800 MHz and 1900 MHz bands. To improve accuracy multiple knife-edge diffraction calculation is applied. For example, the Deygout method can be used for up to three edges (see Figure 5.2). The procedure goes as follows.[3]

- Find the obstacle giving highest loss
- Find the other obstacles (up to two) of each sub-path and calculate losses
- Calculate the total loss as a sum of all the calculated losses.

Depending on the geometry of terrain even Deygout may not give reasonable results. In such cases some other methods must be considered.

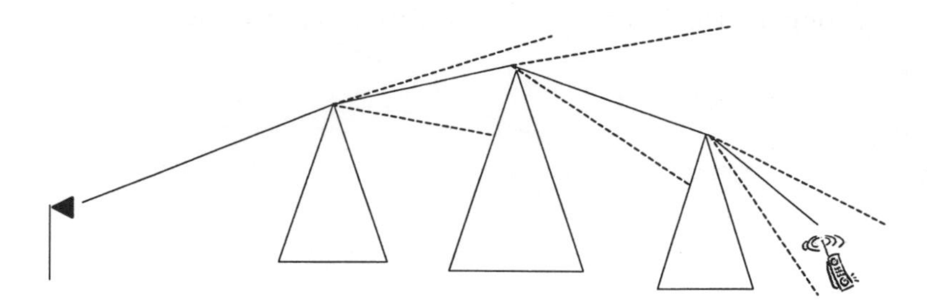

Figure 5.2 Deygout method.

5.2 Macro level models

The propagation models which are commonly used for macro cells are Okumura–Hata [4] and Juul–Nyholm.[5] These models are developed by combining propagation theory and extensive measurement campaigns. The models take several parameters like effective antenna height, terrain type (morphology), and terrain height (topography), frequency, EIRP, etc. These two models are macro cell models which have limitations in terms of frequency, calculation ranges, and base station antenna height.

5.2.1 Okumura–Hata

The Okumura–Hata model gives loss between transmitter and receivers by Equation 5.9.

$$L = A + B\log_{10} f - 13.82\log_{10} h_{BTS} - a(h_{MS}) + 44.9 - 6.55\log_{10} h_{BTS})\log_{10} d + C_m \qquad \text{Equation 5.9}$$

Where a_m is defined as:

$$a(h_{MS}) = (1.1\log_{10} f - 0.7)h_{MS} - (1.56\log_{10} f - 0.8) \quad (1)$$
$$a(h_{MS}) = 3.2(\log_{10}(11.75h_{MS}))^2 - 4.97 \quad (2) \qquad \text{Equation 5.10}$$

In Equation 5.10 the formula (1) is used for small and medium size cities and the formula (2) for large cities.

In Equation 5.9 and Equation 5.10 the definitions used are

L	Path loss (dB)
A	Constant (see Table 5.1)
B	Constant (see Table 5.1)
f	Frequency (MHz)
h_{BTS}	Base station effective antenna height (m)
h_{MS}	Mobile antenna height (m)
d	Distance between mobile and base station (km)
C_m	Area type correction factor.

The value of C_m depends upon the environment. In urban areas C_m is usually above 0 dB but in rural areas the value can even be below –15 dB.

In Equation 5.9 the constants A and B are frequency dependent. The values for these constants are given in Table 5.1.

Table 5.1. The default values for constants A and B in Okumura–Hata propagation model.

	150 1000 MHz	1500 2000 MHz
A	69.55	46.3
B	26.16	33.9

The Okumura–Hata model has several limitations as shown;

- Frequency range 150–1000 MHz and 1500–2000 MHz
- Transmitter antenna height between 30 and 200 meters. (This limitation is often given. However, the Okumura–Hata model does not work well if the average building height is close or higher than the base station antenna height.)
- Calculation range 1–20 km.

5.2.2 Juul–Nyholm

The Juul–Nyholm model predicts field strength in dBμV/m instead of path loss in decibels. Equation 5.11 presents the propagation formula.[5]

$$F = 10\log_{10} P - k_0 + k_1 \log_{10} h_{BTS} - (k_2 - k_3 \log_{10} h_{BTS})\log_{10} d$$

Equation 5.11

In Equation 5.11 the following notations are used

F	Field strength (dBμV/m)
P	Equivalent radiated power (ERP) (W)
$k_0,\ldots, k_3$	Coefficient (see Table 5.2)
h_{BTS}	Base station antenna height (m)
d	Distance between mobile and base station (km).

Table 5.2 Default parameters for the Juul–Nyholm Model for 1800 MHz.

	Range 1–20 km	Range 20–100 km
k_0	23	30
k_1	13	27
k_2	47	50
k_3	–7	3.6

The limitations of Juul–Nyholm are

- Frequency range 800–1000 MHz and 1700–2000 MHz,
- Transmitter antenna height between 30 and 150 m,
- Calculation range 1–100 km.

Juul–Nyholm is a useful model for large cells having a cell radius of over 10 km. Contrary to the Okumara–Hata model the Juul–Nyholm model can be tuned to predict accurately for close proximity to the base station as well from a distance, with the help of the coefficients $k_0,\ldots,k_3$. There are two remarks that must be made concerning the coefficients. First, the coefficients must be tuned separately for 1–20 km and 20–100 km cell ranges. Second, the coefficients are frequency dependent i.e. the coefficients must be tuned as the frequency band is changed.

5.2.3 Antenna height

In cell range calculations the definition of the antenna height may substantially improve the accuracy of the results. There are at least three methods to define the effective antenna height of a base station. If the terrain is flat then the effective antenna height is the same as the antenna height from the ground. However, if the terrain is undulating, the selection of the calculation method may have an impact on the results.

It is also possible to define the base station effective antenna height as a antenna height from the ground without taking into account the terrain height or the location of the mobile.

The method suggested by CCIR defines the effective antenna height as a sum of the height of the base station antenna and the height difference of the ground between the base station and the mobile station. The difference is added only if it is positive (see Figure 5.3).[6]

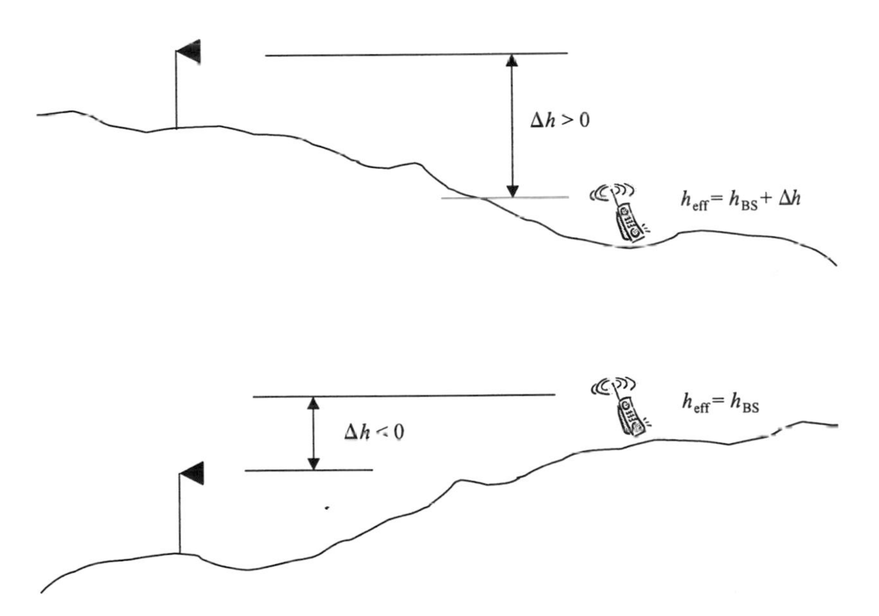

Figure 5.3. Base station effective antenna height calculation using the method suggested by CCIR.

The alternative method is based on the mean height of the ray line above the ground. The effective antenna height is twice the area between the ray and ground profile divided by the distance. In case there are obstructions, a straight is replaced by a series of connected rays touching the obstructions (see Figure 5.4).[7]

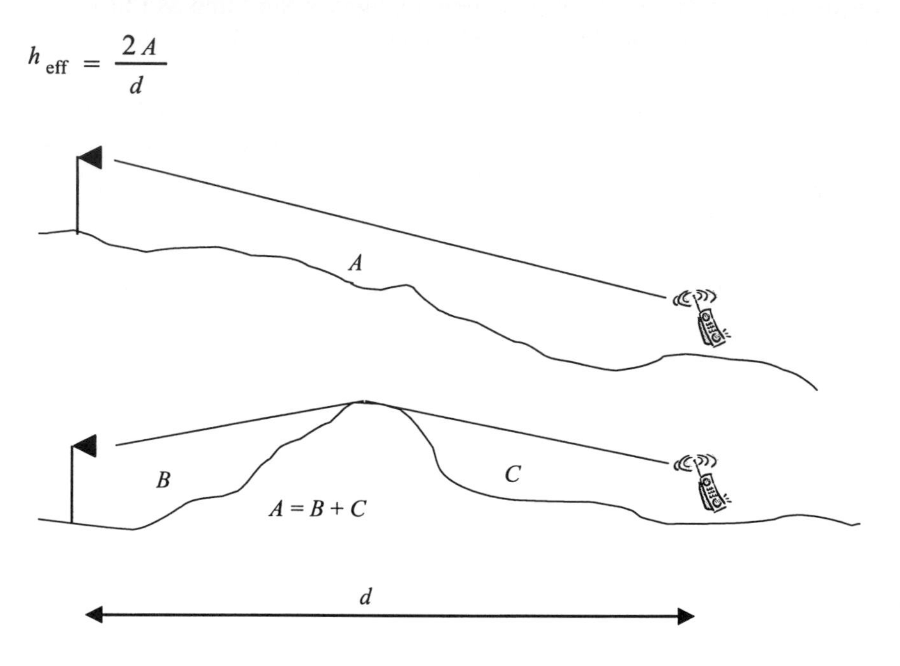

Figure 5.4. Alternative method for calculating the effective antenna height. The distance between the transmitter and receiver is marked with a *d*.

5.3 Propagation measurements and model tuning

Earlier propagation predictions were made based on static propagation models or sets of propagation curves, and terrain data, that is morphology or topology, were not used very effectively. Nowadays it is possible to use computers and digital maps to enhance the predictions.

Before the digital maps and advanced propagation prediction tools can be used, the propagation model should be tuned.

5.3.1 Model tuning measurements

The outcome of model tuning measurements is data that can be used in propagation model tuning. Usually propagation measurements are only carried out when starting to plan a new network, if there is an area with changes in the propagation environment, due to new buildings or roads or if a new frequency band is taken into a use.

Model tuning measurements require a good measurement system, a well-prepared measurement plan, and a lot of experience. The amount of measurements depends on the resolution of the digital map and the size of the target area. If the resolution of the digital map is very high or the target area is large, more measurements should be carried out. An example of measurement equipment is presented in Figure 5.5.

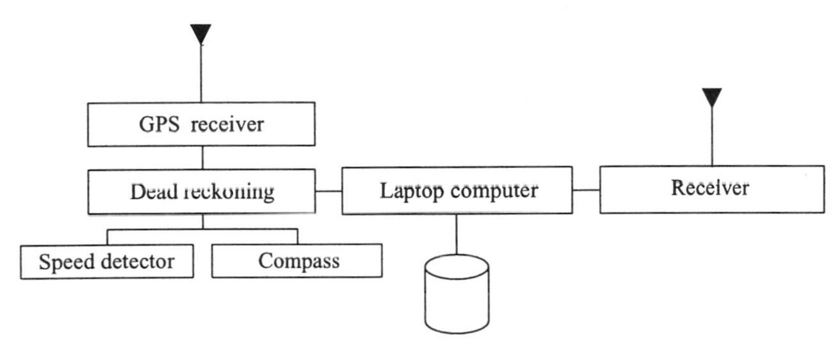

Figure 5.5. A block diagram of a measurement system.

A laptop computer controls the measurement system, collects the measurements and stores the results. The RF receiver collects samples of the measured signal. The receiver should be accurate enough for the measurements (accuracy can be improved by calibrating the receiver and following the instructions provided by the manufacturer). Typically the accuracy of a professional receiver is around ± 1 dB. The use of mobiles in the measurements should be considered carefully as the accuracy of a mobile station is typically ± 4 dB but with calibration the accuracy can be improved to ± 2 dB. A GPS receiver is used to collect location information. The accuracy of GPS is approximately 200–300 m due to selective availability (SA). Recently SA was switched off and the accuracy of GPS improved dramatically. A speed detector and compass are used to improve accuracy in the areas where GPS is not available, as typically with urban shadowing it is possible to lose the GPS signal, at which time the speed detector and compass will help in tracking the route until GPS is available again.

If there are the existing cells available they can be measured for model tuning if all the relevant parameters are known. Basically this means that EIRP, antenna type, direction and height must be known. If there are no existing sites available or the required parameters are not known, a test transmitter can be used instead. It should be noted that the measurement bandwidth should be matched with the 3 dB bandwidth of the transmitted signal. If, for example, the transmitter sends a 25 kHz CW signal, the measurement bandwidth should be around 10 kHz to 15 kHz. If 200 kHz signal is used (GSM) the measurement bandwidth could be, for example, 100 kHz. The exact bandwidths are not given because the receiver may only have a few bandwidth selections available, i.e. the measurement bandwidth cannot be tuned without steps. A block diagram of the test transmitter is shown in Figure 5.6.

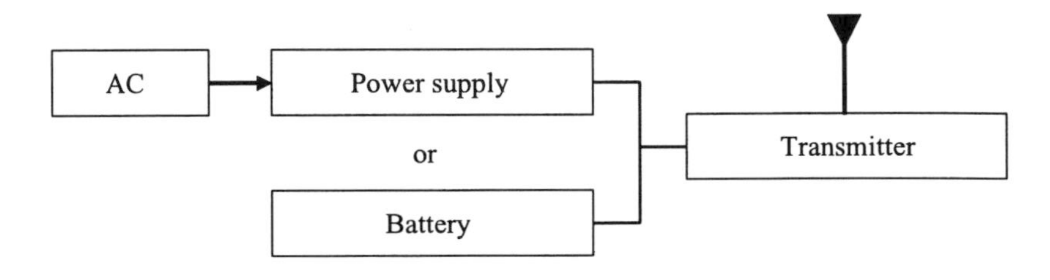

Figure 5.6. A block diagram of a test transmitter.

A measurement system for indoor measurements should at least have a receiver and a computer to collect the measurement data. There are indoor measurement systems available that are integrated with a mobile station. Often the problem with indoor measurement is the lack of location information. This makes it difficult to locate the measurements in the analysing phase. The solution can be, for example, infrared transmitters and an infrared receiver.

5.3.2 Model tuning process

With the utilisation of a digital map, antenna pattern database and the propagation models implemented in a planning system, the propagation models are tuned by matching the predictions with the measured data. Figure 5.7 presents the main steps in propagation model tuning in a planning system. Site and cell data, a digital map and system information (e.g. frequency) are used.

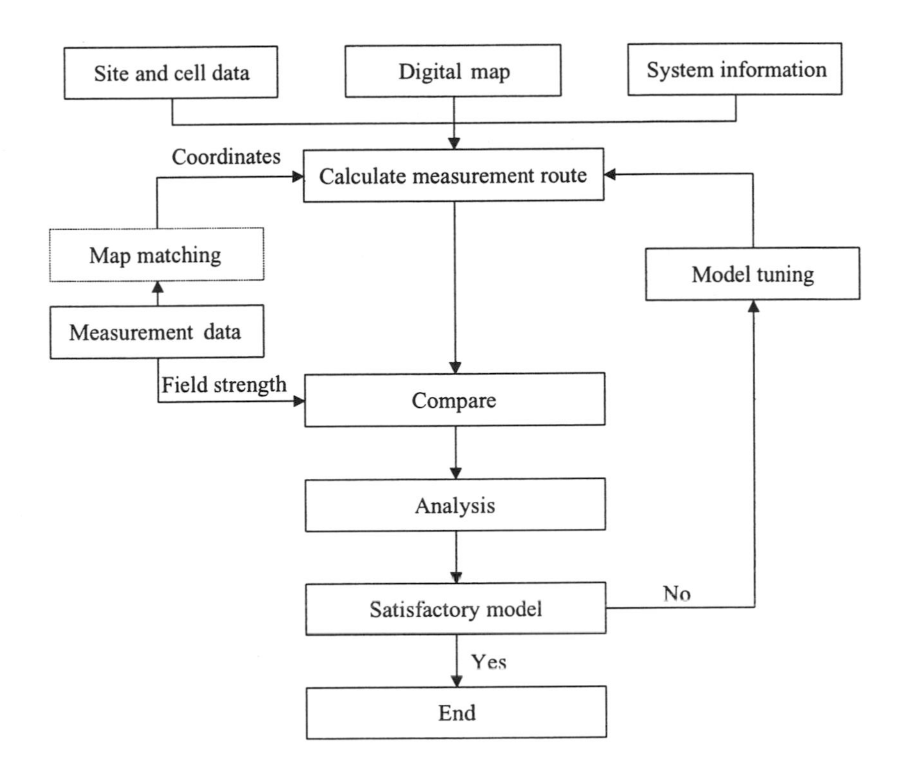

Figure 5.7. Propagation model tuning in a planning system.

When using model tuning measurements for tuning the propagation model, some aspects should be kept in mind. It is important that the prediction is as accurate as possible. It is also important that the measured signal is interference-free and the signal is measured in an area that is neither too close nor too far from the transmitter. To ensure correct measurements these rules should be followed:

- measure only interference-free frequencies
- measure only the main lobe of the transmitting antenna
- avoid or erase line-of-sight (LOS) measurement points (if using NLOS propagation model)
- use differential GPS if possible or match the coordinates with the map
- check coordinate conversion parameters

- measure all the cable losses (both in the transmitting and receiving end)
- measure the output power of the transmitter
- check transmitter antenna installation and ensure that there are no obstacles nearby
- document the measurements carefully.

If the measured frequencies are not interference-free the field strength will be biased; where the required signal is weak due to shadowing or fast fading and the simultaneously interfering signal is strong. This problem arises when measuring a live network because the frequencies are used often. The best solution is to use a dedicated channel for the model tuning measurements. Unfortunately this can be difficult during daytime and therefore it may be more convenient to carry out the measurements at night.

The radiation pattern of an antenna has a main lobe and several side lobes. The main lobe usually has a smooth form without rapid changes in gain. Also, in the main direction, the gain of the antenna is accurate. In dips between the main lobe and side lobes the changes in gain are rapid. Accuracy of the radiation pattern in these dips is poor and thus may cause inaccuracy in the predictions. For this reason it is better to avoid measurements in the areas that are in side lobes or in the back lobe of the transmitting antenna.

Propagation models such as Okumura–Hata and Juul–Nyholm are predicting field strength in non-line-of-sight. This means that they cannot handle situations where there is a line-of-sight. During the measurements it is often difficult to avoid line-of-sight locations, however line-of-sight measurements can be removed after recording. Some tools can also be used to remove line-of-sight measurements with the help of database applications.

The location of the receiver is measured at the same time as the field strength of the transmitter is recorded. Almost always a GPS receiver is used to record the geographical coordinates of the receiver. GPS updates a location once or twice a second and the coordinates are stored in the computer. In differential GPS the GPS receiver is using a correction which tells the receiver how it should correct the error caused by SA. The correction can be calculated using a differential GPS receiver which is located, for example, on the roof of a building where the known location is entered into the receiver. When a differential GPS receiver gets the coordinates of its own location, including the error caused by SA, it can calculate the difference between the correct and the received location. The difference is a vector that is sent to the GPS receiver connected to the measurement system. In some countries differential GPS correction is sent with the help of RDS in FM radio.

Basically all planning systems use an internal orthogonal (or *X–Y*) coordinate system. It may be required to convert geographical coordinates to an orthogonal coordinate system.

For model tuning all the losses, gains and powers must be known. If these are not known the model tuning will give biased results. Cable losses and output power must be measured. Antenna gains and radiation patterns should be correct. Special attention should also be paid to the installation of the antennae.

The documentation of measurements will help in the future in analysing the tuning. Model tuning measurements can also be used for various other purposes in which clear and detailed documentation aids the usage of the measurements.

5.3.3 Model tuning in a planning system

Model tuning is a process in which a theoretical propagation model is tuned with the help of measured values. The models have several parameters that can be changed according to need. The aim is to get the predicted field strength as close as possible to the measured field strength.

5.3.3.1 Measurement import

The first task having confirmed the BS parameters is to import model tuning measurements into the planning system. Depending on the planning system, the measurements are linked to the existing sites and cells by using CELL ID, BTS ID or some other identification. Manual linking is also possible.

Model tuning starts from the selection of the propagation model. For macro level models there are several choices but in principle almost all the models follow the same principles. In this chapter the Okumura–Hata model is used as an example, however, the process can be applied to other models, as well.

After importing the results into the planning system the alignment with the digital map should be checked, observant for GPS SA or inaccuracies in the coordinate conversion parameters. In Figure 5.8 each square represents one measurement. The measurement points line should follow the measured road, though as can be seen, the measurement points have shifted north.

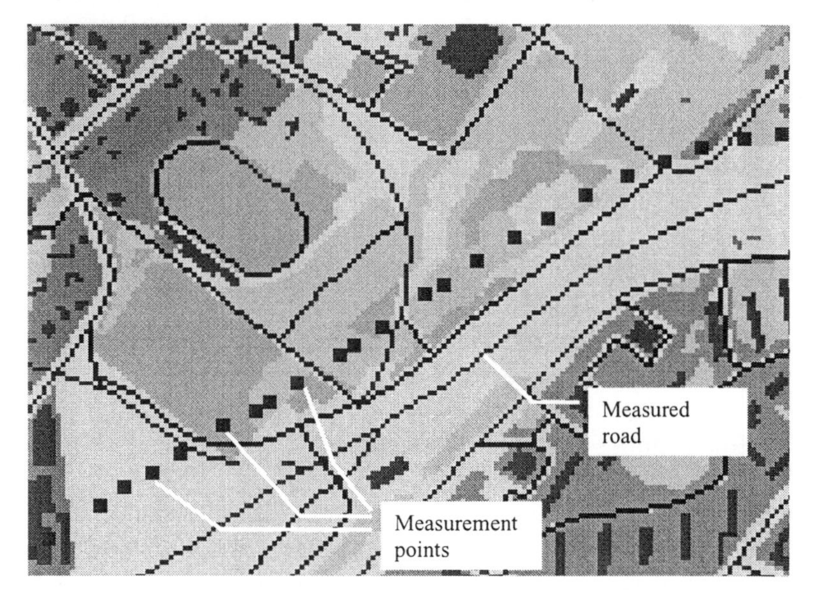

Figure 5.8. Shifted measurement points.

The problem can be solved by shifting the measurements points, as the error in GPS within a short time period is almost constant. The same correction can be used for the entire measurement file.

5.3.3.2 Basic path loss model

Model tuning continues with selecting the model and setting the main parameters. First, the model is selected and its common parameters are set. Usually there are several macro level models in planning systems such as Okumura–Hata, Walfish–Ikegami and Juul–Nyholm. The Walfish–Ikegami model is often said to be a micro cell model but it is more suitable for small macro cells. Of the propagation models mentioned, there can be several correction methods applied to the basic path loss model based on morphology, topology, street orientation, etc.

In Okumura–Hata there are several parameters that can be tuned, however, in some planning systems not all of the parameters may be tuneable, which may cause difficulties in some situations. Usually, in Okumura–Hata model, the slope can be tuned by changing factor C. The effect of C can be seen in Equation 5.12. The first three terms in loss function are independent on the distance d. Because log(d) has a coefficient $C - 6.55x\ \log_{10}(h)$, by changing C the slope of the model can be tuned. In the rural environments C is lower and in urban areas it becomes higher. The city type environment affects the correction that is made based on the mobile antenna height. If the mobile antenna height is 1.5 m (commonly used in network planning), the selection does not change the correction (note that in Equation 5.12 the correction term $a(h_{MS})$ is left out because with the mobile antenna height 1.5 m, the mobile antenna height correction is 0 dB). Correction D affects the effect of antenna height. The Okumura–Hata model is suitable in cells that have antennae installed well above roof-tops. In those cases the default value of 13.82 dB is suitable. If antennae are installed close to the average roof-top level by changing parameter D the accuracy of predictions can be improved even if the basic model is no longer working well.

$$L = A + B\log_{10} f - D\log_{10} h_{BTS} + (C - 6.55\log_{10} h_{BTS})\log_{10} d \qquad \textit{Equation 5.12}$$

In the Juul–Nyholm propagation model the parameters are defined separately for the distances 1–20 km and 20–100 km. The advantage in Juul–Nyholm is the calculation range and possibility to tune the model for very large cells. However, when giving parameters k_0, ..., k_3 for 1–20 km and 20–100 km, it is possible to have a point of discontinuity in the prediction at the distance of 20 km.

5.3.3.3 Morphology corrections

The correction due to morphology is the best known correction method for the common propagation models. The result of the basic propagation model is adjusted according to the terrain types between the mobile station and the base station. The example shows how the correction is usually calculated.

Example: Morphology correction

The distance between the base station and the mobile station is 1.5 km. On the digital map there are 30 pixels (50 m x 50 m) between the base station and the mobile. Each pixel presents the terrain type within the 50 m x 50 m area.

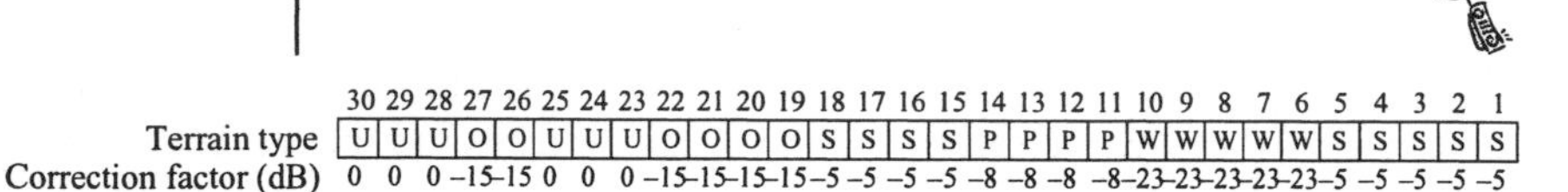

Pixel size: 50 m x 50 m

Figure 5.9. Example of terrain type corrections. The following notations are used: U = urban, S = suburban, P = park, O = open and W = water.

The morphology correction is calculated as an average of the pixels between the mobile station and base station. In Figure 5.9 there are 6 urban pixel, 9 suburban pixels, 4 park pixels, 6 open pixels and 5 water pixels between the mobile and the base station. The average of the correction factors in this example is –9.4 dB. The basic propagation model is corrected by adding the calculated correction to the prediction result (see C_m in Equation 5.9).

In the example above there are a few remarks that must be made. The distance between the base station and mobile was 1.5 km. It is not very likely that the area close to the base station has a great impact on the received power of the mobile station; the area close to the mobile are more important for the prediction, thus there are ways to weight the area close to the mobile station. First, the calculation distance can be shorter than the distance between the mobile station and the base station. This means that only the pixels close to the mobile stations are considered when calculating the correction. If in the previous example the calculation distance is changed from 1.5 km down to 500 m the average of the correction factors is –14 dB. In this case the difference between the corrections is 4.6 dB which means that the selection of the calculation distance is important. In the example all the terrain type pixels within the calculation distance had the same weights but the calculation distance was changed. In the second example the pixels closer to the mobile station will have greater impact on the correction made to the prediction because the weights for correction factors are introduced. In Figure 5.10 the calculation distance and weights are presented.

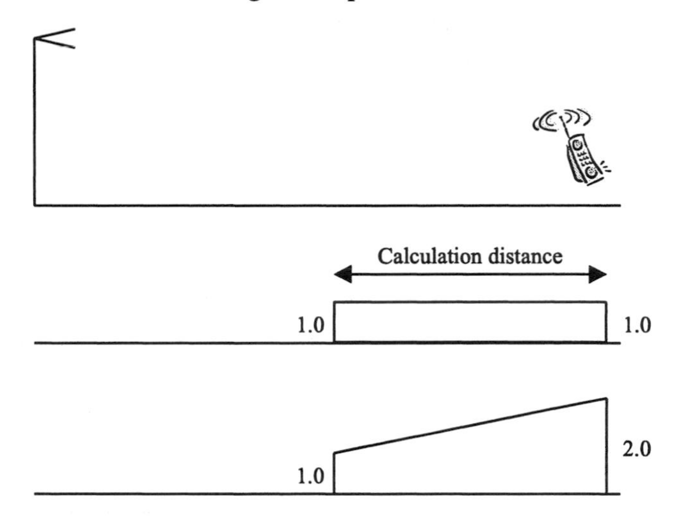

Figure 5.10. Calculation distance and weighting of the correction factors.

If the correction for the basic propagation model is calculated based on the '1.0–2.0' weights as shown in Figure 5.10 the result is –12.33 dB and if the calculation distance is kept as 500 m (see Figure 5.11).

	10	9	8	7	6	5	4	3	2	1
Terrain type	W	W	W	W	W	S	S	S	S	S
Correction factor (dB)	–23	–23	–23	–23	–23	–5	–5	–5	–5	–5
Weights	1	1.11	1.22	1.33	1.44	1.56	1.67	1.78	1.89	2
Normalised weights	0.67	0.74	0.81	0.89	0.96	1.04	1.11	1.19	1.26	1.33
Normalised correction factors	–15	–17	–19	–20	–22	–5.2	–5.6	–5.9	–6.3	–6.7

Figure 5.11. Linear weights for terrain type correction factors (example). The average of the normalised correction factors is –12.33 dB. Notations are the same as used in Figure 5.9.

It is possible to adjust the calculation distance automatically cell by cell based on the average arrival angle of the received signal and Fresnell zones. This method does not always give reliable results because of the environment, but more often because of the resolution of the digital map.

5.3.3.4 Topographic correction

Topographic corrections in coverage predictions are usually done with the help of the terrain height model. In many planning systems the correction can be made even if the terrain height information is not available by using, for example, Okumura method.[8]

Clearance Angle method can be useful in cases where, for example, diffraction method is not working well. In clearance angle method the loss is calculated based on the angle between the horizontal plane and the direction of the base station antenna.[6]

The definition clearance angle is presented in Figure 5.12.

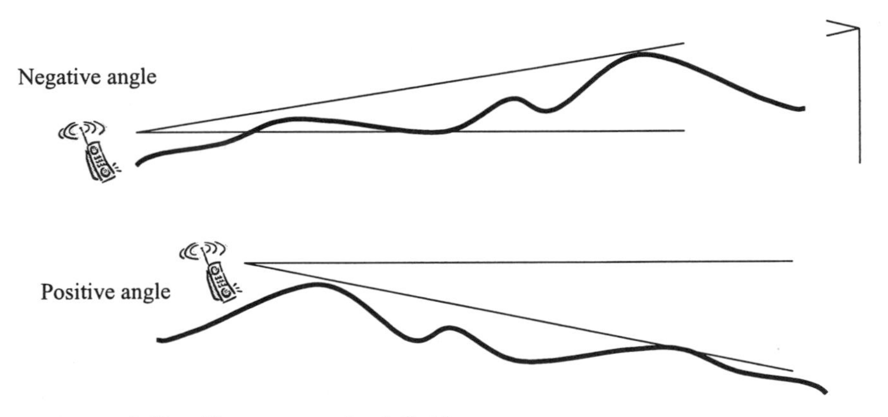

Figure 5.12. Clearance angle definition.

The correction is calculated based on the clearance angle. An example of a correction function is presented in Figure 5.13 which shows that the correction is roughly between –8 dB and 3 dB depending the clearance angle. In some planning systems the correction curve is predefined, which limits the usage of the clearance angle method in some cases.

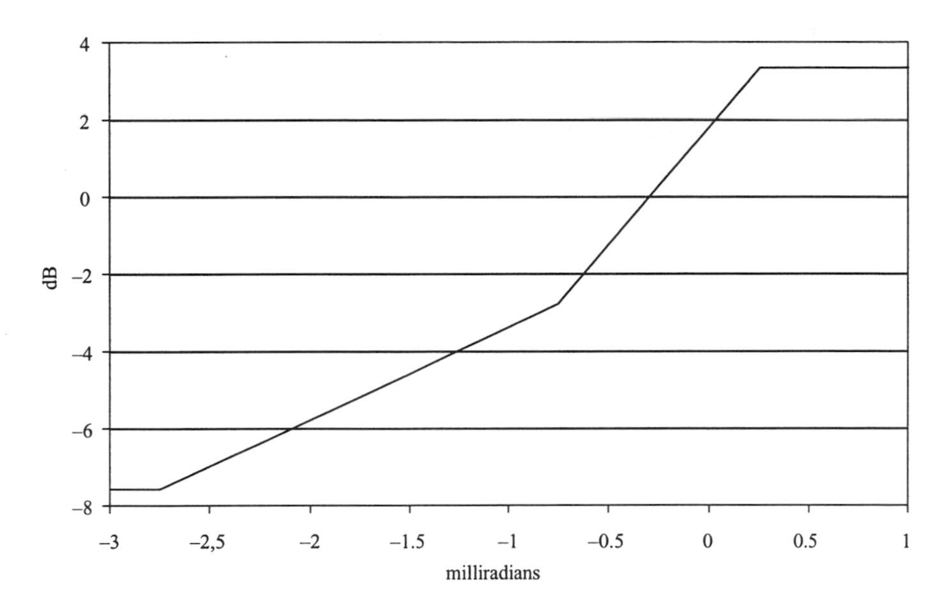

Figure 5.13. Clearance angle correction curve.

Diffraction method is the last topography correction method presented here. To calculate the loss due to diffraction it is possible to use either the knife-edge method or Deygout method presented in 5.1.1.

5.3.3.5 Street orientation

The last correction method for the basic model is *street orientation correction* (see Figure 5.14). This correction was used for the Walfish–Ikegami model but the same correction can be applied to any other model as well. The correction function is predefined.[9]

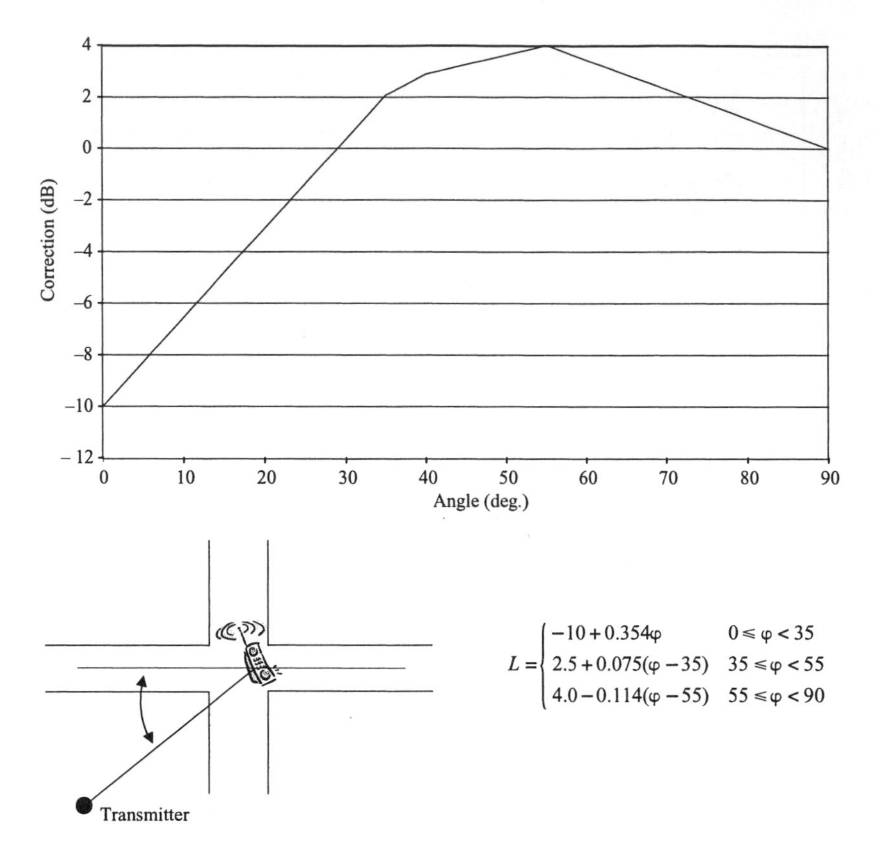

Figure 5.14. Correction function for street orientation.

5.3.3.6 Tuning of morphology correction parameters

When some default values for the basic path loss model and topographic correction are set it is possible to start tuning morphology correction factors.

The aim in morphology correction factor tuning is to find such correction factors for each morphology class that the predicted field strength differ as little as possible from the measured field strength. There are two measures to validate the tuning. First, the mean value of the prediction should be the same as the mean value of the measured values. Second, the standard deviation of the difference between the measured and predicted values is calculated. The lower the deviation the better the model tuning is. In the model tuning both of these measures should be considered. If the mean value of the difference is not close to zero there is one or several systematic errors either in the measurements or in predictions. If the standard deviation is large the basic path loss model, topographic correction or the correction factors for morphology correction are wrong. This should be a sign for the next iteration round.

The goodness of tuning can be monitored in various ways. Figure 5.15 shows the measured and predicted field strengths as two separate curves. The x axis is the number of measurement points so the window shows, for example, how the field strength changes during the measurements. In the example there are a few LOS measurements that should be removed before the model tuning.

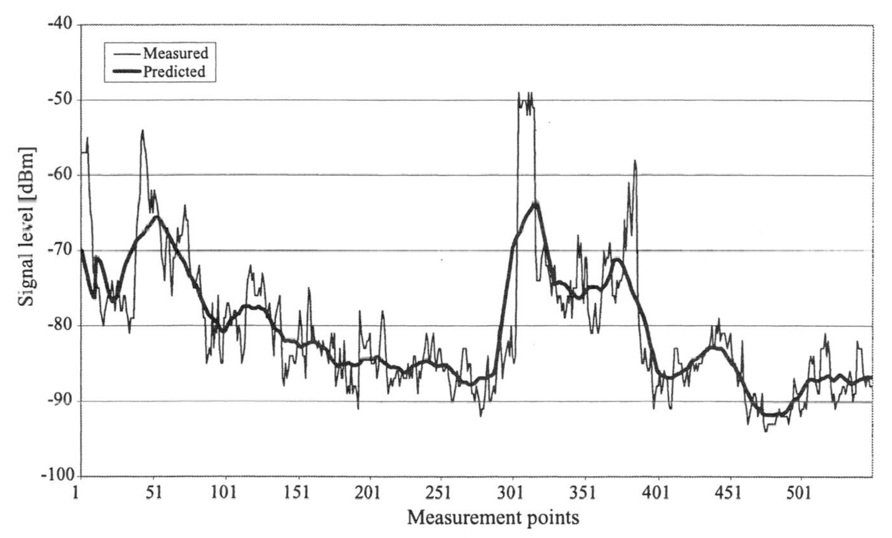

Figure 5.15 Measured and predicted signal level values.

The histogram (Figure 5.16) shows the distribution of the difference between the predicted and measured field strength. The results are good if the mean value of the distribution is close to zero and the distribution is as sharp as possible.

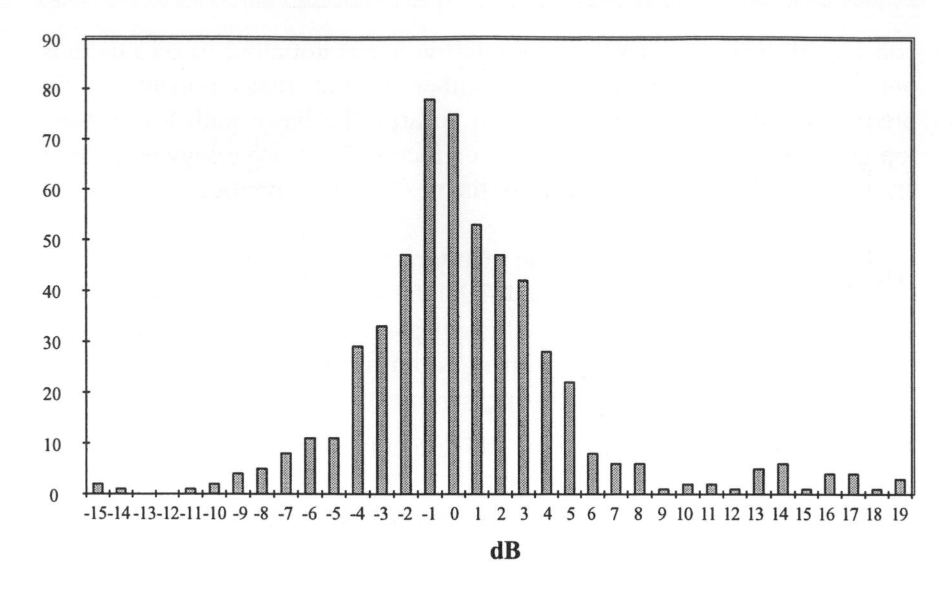

Figure 5.16. Histogram of the difference between the measured and predicted signal levels.

Figure 5.17 shows the lines that are fitted to the measured points and the predicted points. The line is fitted based on the distance (x) between the receiver and the transmitter. The form of the line is $a + b\ \log_{10}(d)$, where the letter b presents the slope of the line.

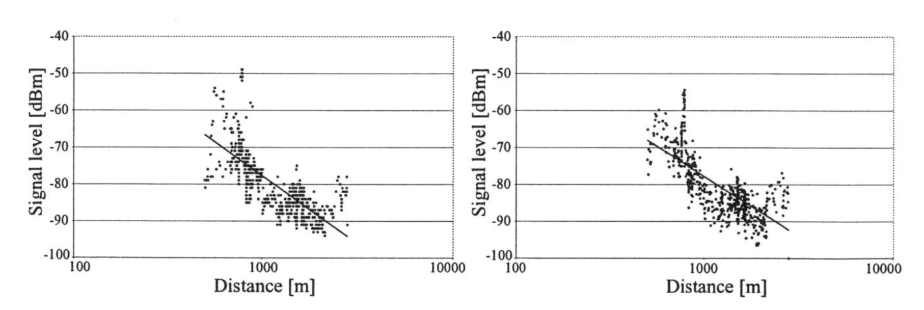

Figure 5.17. Fitted lines for the measured and predicted signal levels.

The morphology correction is changed either semiautomatically or manually. The semiautomatic method does not always give realistic results and thus manual tuning is needed in order to improve the results.

Figure 5.18 shows an example of a semiautomatically tune cell.

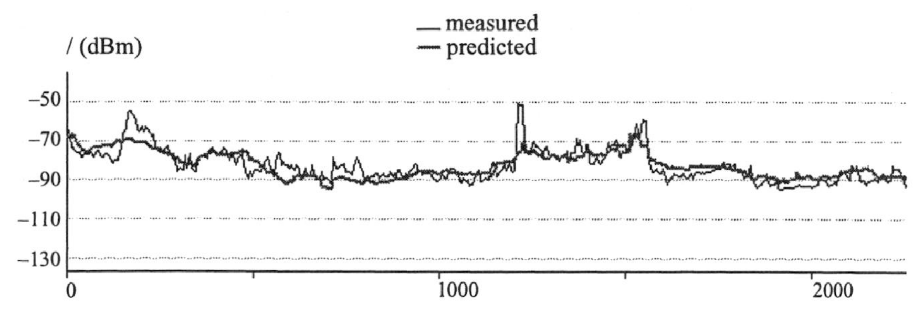

Figure 5.18. The results of the morphology correction factor tuning.

The tuning for this cell is good because the mean value is very close to zero. The standard deviation of the error between the predicted and measured field strengths is also quite low. The result could be improved further by removing the LOS measurement points because the building database was not in use; thus there is no way to identify LOS measurement points.

The process of model tuning can be presented with two flow charts. First, the basic path loss model is tuned according to the process presented in Figure 5.19.

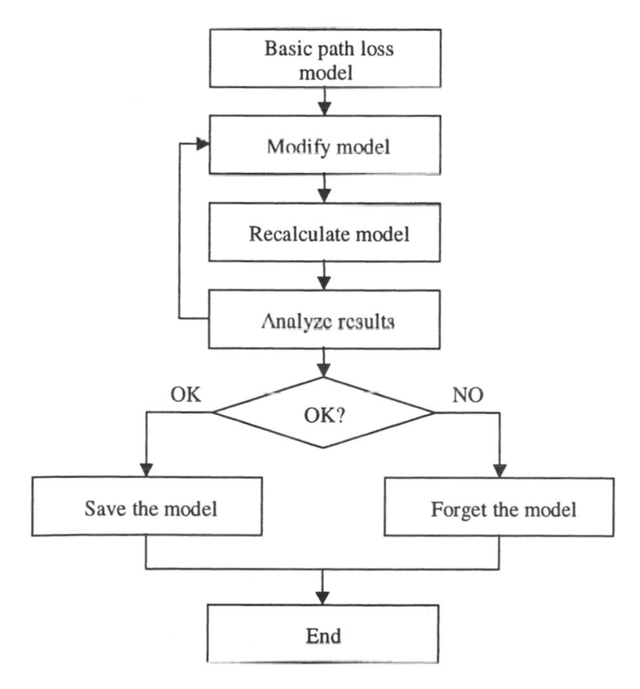

Figure 5.19. Basic path loss model tuning process.

As Figure 5.19 shows the process is iterative. Because there are several parameters to be tuned in the basic path loss model, the tuning takes usually several rounds. Also, the result of the tuning is compromising between accuracy and the resources—one can always improve the prediction by fine-tuning the model but one has to accept some errors in prediction if the time reserved for tuning is limited.

The process for correction factor tuning is shown in Figure 5.20. The process flow chart starts from the situation in which the measurement results are imported into the planning system. Also, the measurement results are matched with the digital map and line-of-sight locations are removed.

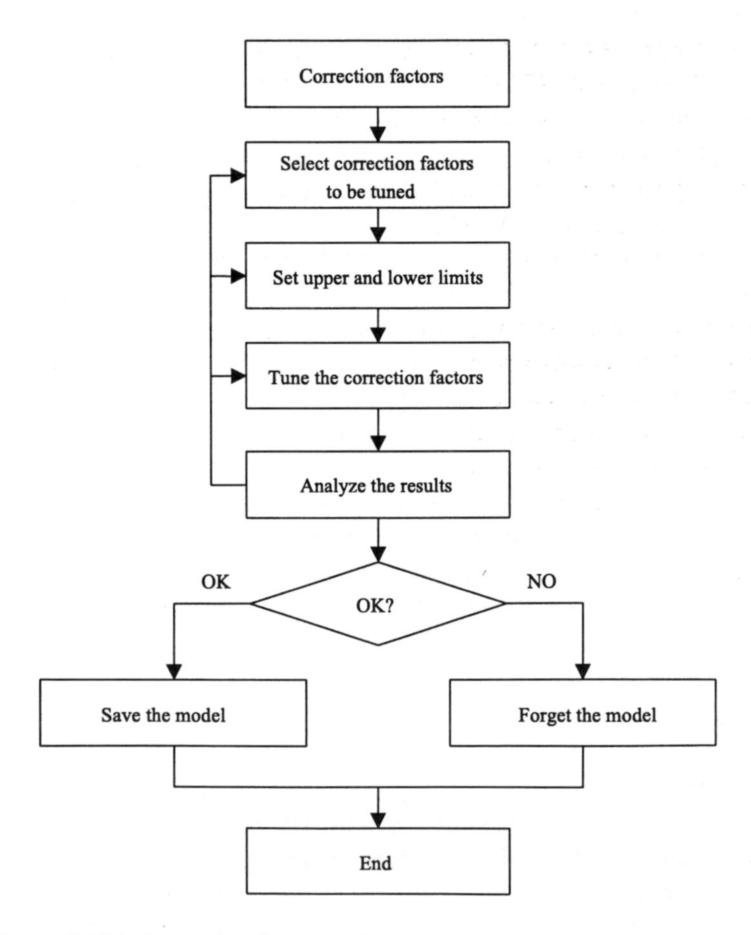

Figure 5.20. Correction factor tuning.

If there are only a few pixels of some morphology classes in the target area it is not practical to tune those classes because the accuracy of the results will not be very good. A rule of thumb could be that the portion of the tuned morphology class should be at least 5 percent of the total amount of the pixels. If there are a very large number of measurement samples available it is possible to tune morphology classes having a smaller portion of the total pixels. For each tuneable morphology class upper and lower limits for the correction factors should be set. By doing this some unrealistic results can be avoided. It may happen that the automatic tuner proposes correction factors that are approaching the limits. If this happens one should check if the limits are reasonable and if there is a need to change the basic propagation model. Usually correction factor tuning requires several rounds of tuning and the basic propagation model may have to be adjusted several time during the process. If everything is done for the tuning but the results are not good the following topics should be checked;

- base station configurations are correct (antenna line, powers, etc.)
- measurement samples are correct, i.e. measurement equipment had the right parameters and settings, coordinates are correct, etc.
- digital map has all the required layers and the accuracy of the map is high enough.

When the propagation model is tuned it should be applied for the cells where it can work. Often the same propagation model is used for the whole network. This is not the optimal situation in most of cases but the model tuning should be carried out for the regions that differ from each other. In theory one propagation model could be enough but there are factors that demand different propagation models for different areas. For example, the classification of terrain types may not be perfect in the digital map because of the limitation of the terrain type classes. Also, in some areas some terrain types may not exist thus the correction factors of these terrain types cannot be tuned.

5.4 Digital maps

A digital map is the basic component of a planning system. All the tasks carried out in a planning system are in some way related to the digital map(s). Obviously, coverage planning is directly related to the digital map because the information on the map is needed for the predictions. Digital maps are also needed in frequency planning, capacity planning, interference analysis, etc. Because the digital maps are one of the basic components in a planning system the selection and the usage of the maps should be considered carefully.

The data in a digital map is presented either in a raster format or in a vector format. Typical raster data are topographic and morphographic data. Other raster data layers can be, for example, building heights and traffic density. Roads, borders and texts are usually in vector format. Also, in the new micro cellular models the buildings are vectorised.

5.4.1 Quality criteria for digital map

Common quality criteria for a digital map are resolution, accuracy, age of the source material, quality of the digitalisation and quality of classification. The resolution means the smallest unit on the map that can be presented. Usually the resolution is expressed as a pixel size. In Figure 5.21 there are two maps loaded simultaneously. The map above is with 5 m resolution and the map below is with 50 m resolutions.

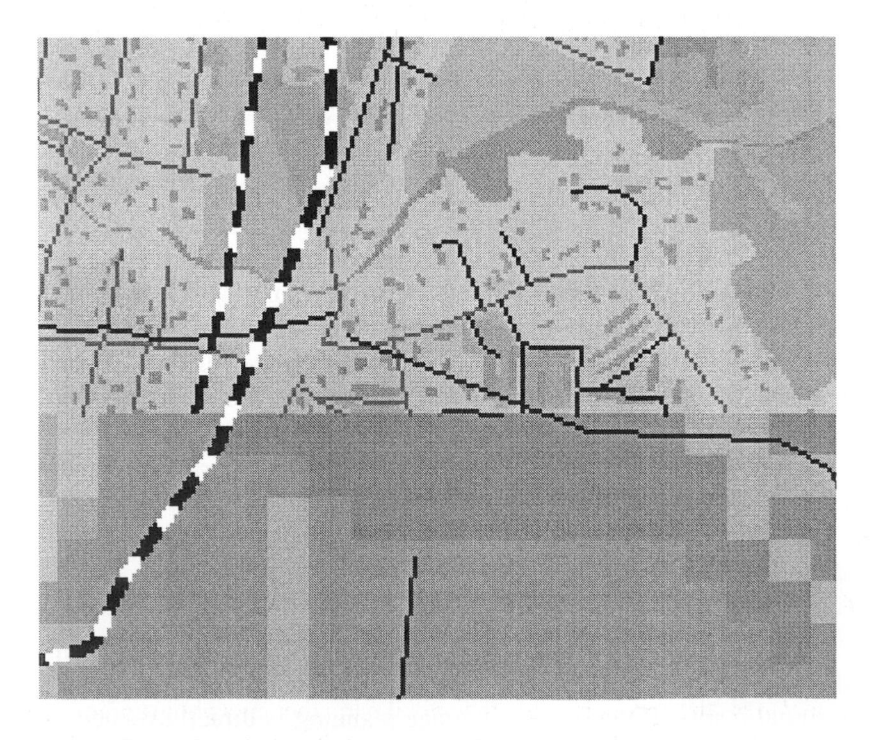

Figure 5.21. Resolution and accuracy of a digital map.

In both maps the accuracy seems to be close to resolution, i.e. both maps are well produced. It may happen that the map has good resolution but the resolution is not fully utilized. In Figure 5.22 the area is loaded with 50 m resolution. If a digital map is produced with lower accuracy than the resolution, the difference is similar than it is in Figure 5.21 and Figure 5.22. As can be seen the details are lost in Figure 5.22.

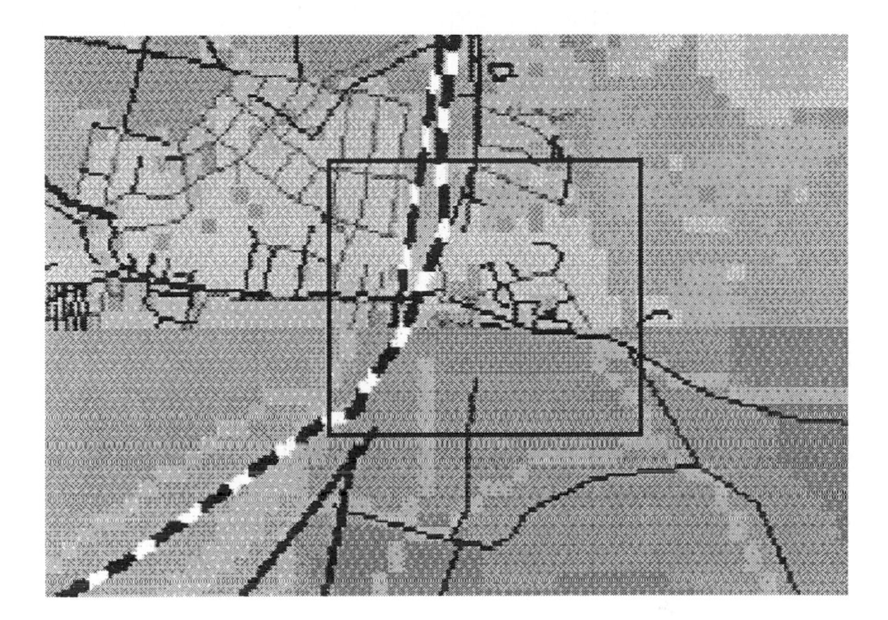

Figure 5.22. 5 m map loaded with 50 m resolution. The area shown in Figure 5.21 is marked with a rectangle.

The quality of digitalisation and classification can be seen in Figure 5.22. The map on the lower part of the figure is produced by using older and less accurate source material. The 5 m map has more accurate classification even if it is loaded with 50 m resolution. The age of material becomes important in the areas where new buildings are constructed or there are other changes.

5.4.2 Raster data

The raster data is presented with pixels. Each pixel has only one value that defines the property of the pixel. In the morphographic layer, different land-usage classes are presented as different terrain-type classes. An example is presented in Figure 5.23.

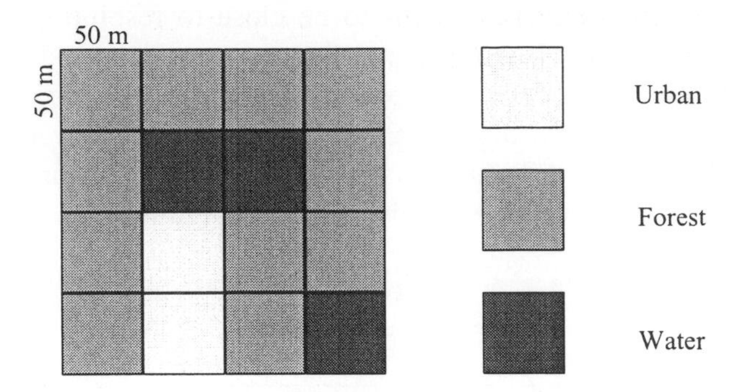

Figure 5.23. Raster morphographic data.

Usually 4 to 10 morphographic classes are used to classify the terrain types. The classification and the number of types needed depends on the terrain itself. In some cases only a few classes are enough but, for example, vegetation may require several classes.

Terrain height is presented in raster format in some planning systems. Each pixel has its own terrain height information as shown in Figure 5.24. The value is calculated based on the height model. If the resolution of the map is poor there can be severe errors in the terrain height model.

50 m (pixel width) × 50 m (pixel height)

54	52	55	56
53	53	54	56
52	54	55	57
53	55	56	58

Figure 5.24. Raster topographic data. The terrain height varies between 52 and 58 m in this example.

In Figure 5.25 the height information for pixels 1 and 3 are easy to determine. Also, the result does not change if the maximum value or average is used. However, with pixel 2 there is a problem with how to select the value. The average 58 m is fairly good for radio network planning but for radio link planning it cannot be used. If the maximum

63 m is selected the predictions will give too pessimistic results because the peak is not blocking the radio waves totally.

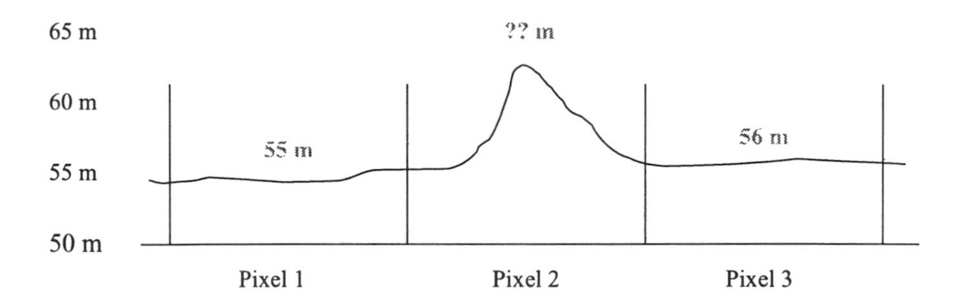

Figure 5.25. Possible problems in topographic layer.

Building heights and traffic density data are expressed in raster format. Each building is presented with pixels having height information. The format is similar to the topographic layer. Traffic data shows the relative traffic within the target area. If all the pixels have the same value the traffic is evenly distributed. The absolute traffic is calculated by assigning a value for the basic unit.

5.4.3 Vector data

In a planning system some of the data is presented in vector format. Typically roads and borders are presented with vectors. Also, text strings can be shown as vectors. The vector data is given with coordinates. Also, some attributes can be defined for the vectors. These attributes can be fonts, colours, line width, etc.

5.5 Micro level models

There are several micro cellular models introduced for small cells. Some models are designed for small cells and some for real micro cells. For example, the Walfish–Ikegami model is said to be a micro cellular model but it is better used for small macro cells in urban areas. The real micro cellular models are not yet in large-scale use.

5.5.1 Walfish–Ikegami

The COST-231-Walfish–Ikegami prediction model is said to be a micro cell model. The parameters for the model are building separation in meters, building average height in meters, roads width in meters and road orientation angle in degrees. These parameters can be defined manually for each cell or they can be taken from the digital map having building height layer.[9]

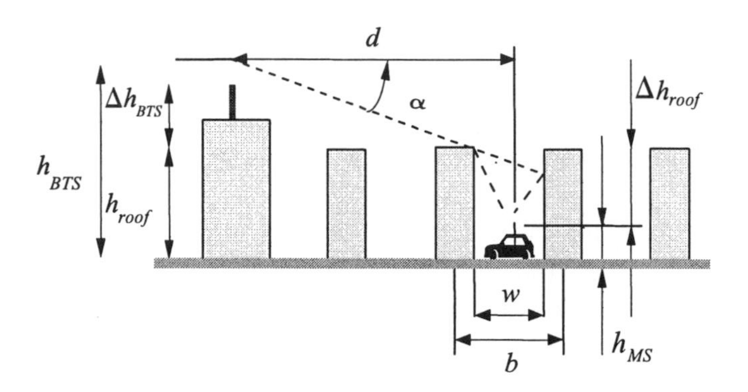

Figure 5.26. Walfish–Ikegami propagation model parameters.

Building separation is the distance between the centres of two buildings (b in Figure 5.26) and the building average height is the average height of all the buildings inside the coverage area of the cell (h_{roof}). Roads width (w) is given in meters and the road orientation angle (φ) is given in degrees (Figure 5.27).

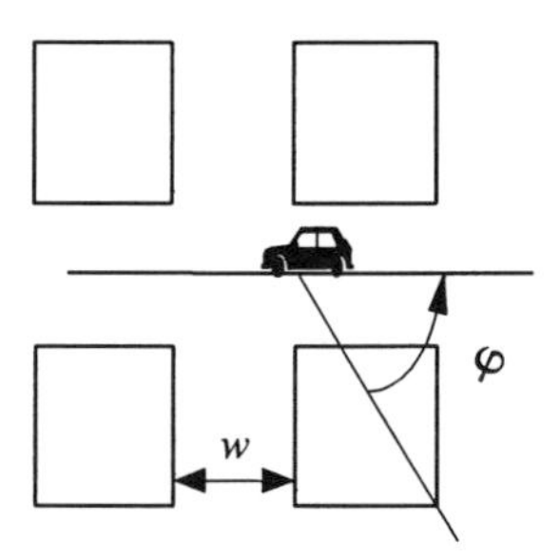

Figure 5.27. Definition of street orientation.

The Walfish–Ikegami has NLOS and LOS propagation formulae. The basic path loss formulae for LOS is

$$L = 42.6 + 26\log_{10}(d) + 20\log_{10}(f)$$ *Equation 5.13*

and for NLOS

$$L = 32.4 + 20\log_{10}(d) + 20\log_{10}(f) + L_{rts} + L_{msd}$$ *Equation 5.14*

In Equation 5.13 and Equation 5.14 the following definitions are used

L	path loss (dB)
f	frequency (MHz)
d	distance (km)
L_{rts}	rooftop-to-street diffraction and scatter loss (dB)
L_{msd}	multi-screen diffraction loss (dB).

The rooftop-to-street diffraction and scatter loss is given in Equation 5.15.

$$L_{rts} = -16.9 - 10\log_{10}(w) - 10\log_{10}(f) - 20\log_{10}(\Delta h_{MS}) - L_{ors}$$ *Equation 5.15*

In Equation 5.15 street orientation correction is defined

$$L_{ors} = \begin{cases} -10 + 0.354\varphi & 0^0 \le \varphi < 35^0 \\ 2.5 + 0.075(\varphi - 35) & 35^0 \le \varphi < 55^0 \\ 4.0 - 0.114(\varphi - 55) & 55^0 \le \varphi < 90^0 \end{cases}$$ *Equation 5.16*

The other definitions used in Equation 5.15 are

$$\Delta h_{MS} = h_{roof} - h_{MS}$$
$$\Delta h_{BTS} = h_{BTS} - h_{roof}$$ *Equation 5.17*

The multi-screen diffraction loss is given in Equation 5.18.

$$L_{msd} = L_{bsh} + k_a + k_d \log_{10}(d) + k_f \log_{10}(f) - 9 \cdot \log_{10}(b) \qquad \text{Equation 5.18}$$

In Equation 5.18 L_{bsh}, k_a, k_d, k_f are defined as

$$L_{bsh} = \begin{cases} -18\log_{10}(1 - \Delta h_{BTS}) & h_{BTS} > h_{roof} \\ 0 & h_{BTS} \le h_{roof} \end{cases} \qquad \text{Equation 5.19}$$

$$k_a = \begin{cases} 54 & h_{BTS} > h_{roof} \\ 54 - 0.8\Delta h_{BTS} & d \ge 0.5km \text{ and } h_{BTS} \le h_{roof} \\ 54 - 0.8\Delta h_{BTS}\dfrac{d}{0.5} & d < 0.5km \text{ and } h_{BTS} \le h_{roof} \end{cases} \qquad \text{Equation 5.20}$$

$$k_d = \begin{cases} 18 & h_{BTS} > h_{roof} \\ 18 - 15\dfrac{\Delta h_{BTS}}{h_{roof}} & h_{BTS} \le h_{roof} \end{cases} \qquad \text{Equation 5.21}$$

$$k_f = \begin{cases} -4 + 0.7\left(\dfrac{f}{925} - 1\right) & \text{for suburban areas} \\ -4 + 1.5\left(\dfrac{f}{925} - 1\right) & \text{for urban areas} \end{cases} \qquad \text{Equation 5.22}$$

The term k_a represents the increase of the path loss for base station antennas below the rooftop of the adjacent buildings. The terms k_d and k_f control the dependence of the multi-screen diffraction loss versus distance and radio frequency, respectively.

The restrictions on the Walfish–Ikegami model are listed in Table 5.4. The Walfish–Ikegami model works well for an antenna height greater than the mean rooftop level, but becomes less accurate for small grazing angles and when the antenna is located below the rooftop level. It is useful for small macro cells in urban areas because other models like the Okumura–Hata do not work well due to short distances.

Table 5.4. Restrictions of Walfish–Ikegami.

Frequency	800–2000 MHz
h_{BTS} antenna height	4–50 m
h_{MS} antenna height	1–3 m
Distance	0.02–5 km

5.5.2 Other microcellular models

A lot of research has been done in the field of propagation in microcellular environment. Some microcellular models have been developed but unfortunately they are usually a compromise between accuracy and calculation time. Currently the models require too much time if they are used for large-scale microcellular network planning.

5.6 Coverage planning parameters

5.6.1 TX parameters

TX parameters are given for each cell separately. These parameters define the EIRP of a cell, antenna direction and the radiation pattern of an antenna.

TX power defines the output power of the base station. Depending on how the value is used together with *Losses* the *TX power* can be after or before the combiner. It may be easier if *TX power* is defined as an output power of a base station at the antenna port. Normally there are losses in the antenna line. Cable losses, power spilter losses, etc., can be added up and one value, *Losses*, can be used to present all the losses between the antenna connector of the base station and the transmitting antenna.

In antenna selection there are three important issues that are interrelated. Antenna gain, horizontal 3 dB-beamwidth and vertical 3 dB-beamwidth. The selection of the antenna type depends on the coverage requirements and interference situation (see Chapter 3). Radiation patterns for a typical GSM antenna is presented in Figure 5.28.

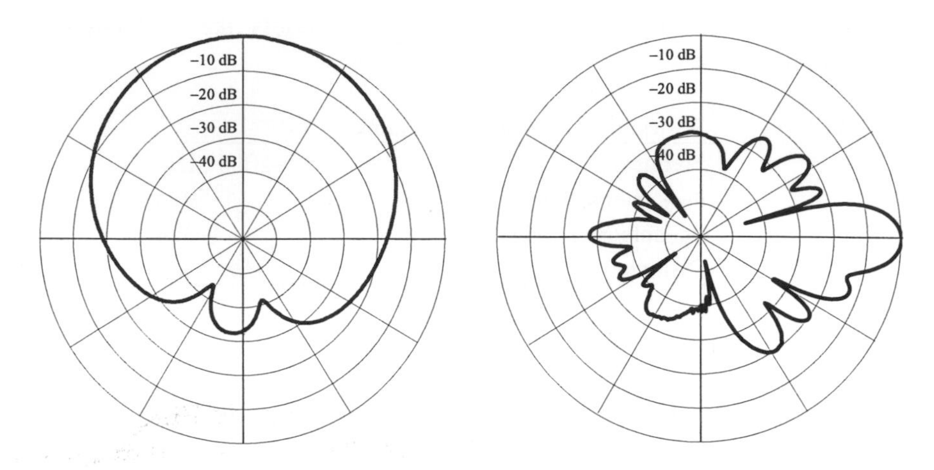

Figure 5.28. Radiation patterns for horizontal and vertical planes.

The gain of the antenna shows how well the radiated power is directed to the desired directions. The selection of horizontal beamwidth depends on the area to be covered. Narrow beam antennae are used when the covered area has a narrow shape. Also, they can be used to limit interference for the cell to the direction that is not in the main coverage area. The selection of vertical beamwidth depends on the environment and the height of the antenna. In rural areas narrow vertical beamwidths are mainly used to achieve the maximum antenna gain. However, in urban areas, if downtilting is used, the vertical beamwidth should be quite narrow, as well. If the vertical beamwidth of the antenna is more than 20° the downtilting angle should be very large before any benefits are gained.

Additional attenuation in the radiation pattern is used to make coverage more realistic in the cases where the antenna is installed, for example, on a wall. In some planning system it is possible to block the back slope thus the coverage and interence analysis will be more accurate. In Figure 5.29 the back slope the radiation pattern is blocked by introducing additional attenuation for the direction between 90° and 270°. If the antenna is installed on the wall the radiation pattern is not exactly as the radiation pattern presented in Figure 5.29 but for most of the cases the presented pattern is accurate enough for coverage and interference analysis.

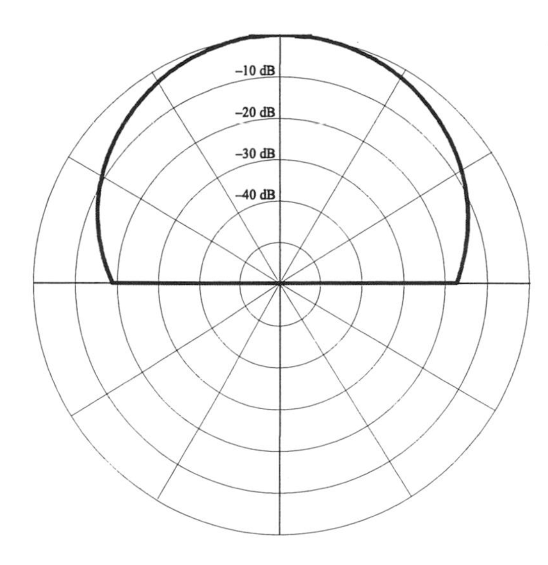

Figure 5.29. Sector attenuation.

Other parameters that have to be given for the antenna are bearing, tilt angle and height. Bearing indicates the direction of the antenna. However, in some countries the declination may cause problems if the antennas are directed with the help of compass because compass course can differ from the direction shown on the map. With tilting it is possible to improve coverage in some cases but usually it is used to improve the *C/I* ratio. The positive tilting angles present downtilting and uptilting is done with negative tilt angles.

5.6.2 Calculation area

Calculation area is very important for coverage calculation but also for frequency planning, interference analysis, etc. When defining calculation regions for cells one must make compromises between calculation time and accuracy. If calculation regions are too large the calculation of coverage takes a long time. On the other hand, too small a calculation region cuts the coverage and influence of the results of, for example, frequency planning. Calculation regions should be selected in such a way that *C/I* or *C/N* ratios will not be affected by too small a calculation region. But the calculation regions do not have to be any larger than necessary to fulfil the criteria.

The impact of too small a calculation region to coverage is presented in Figure 5.30. The coverage areas of the cells are cut in the example. The coverage areas in this case should be much larger.

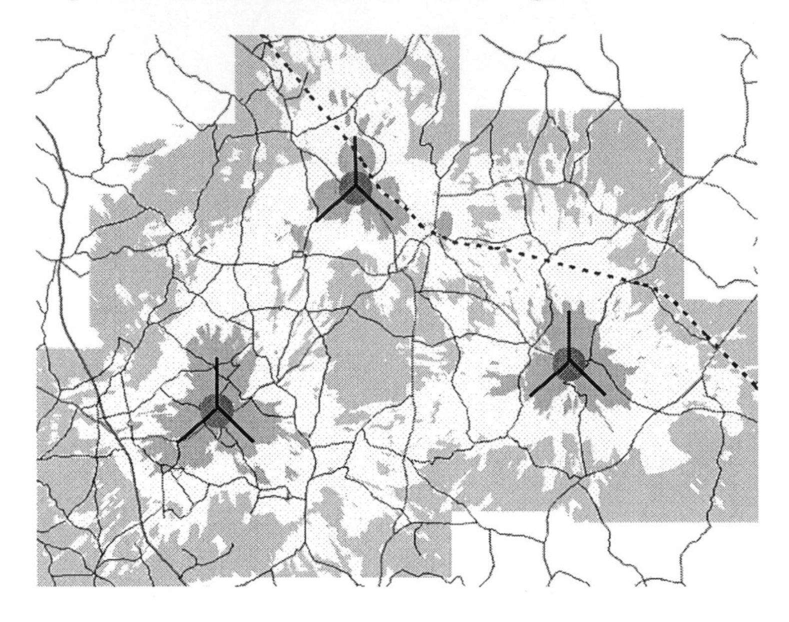

Figure 5.30. Too small a calculation region for coverage.

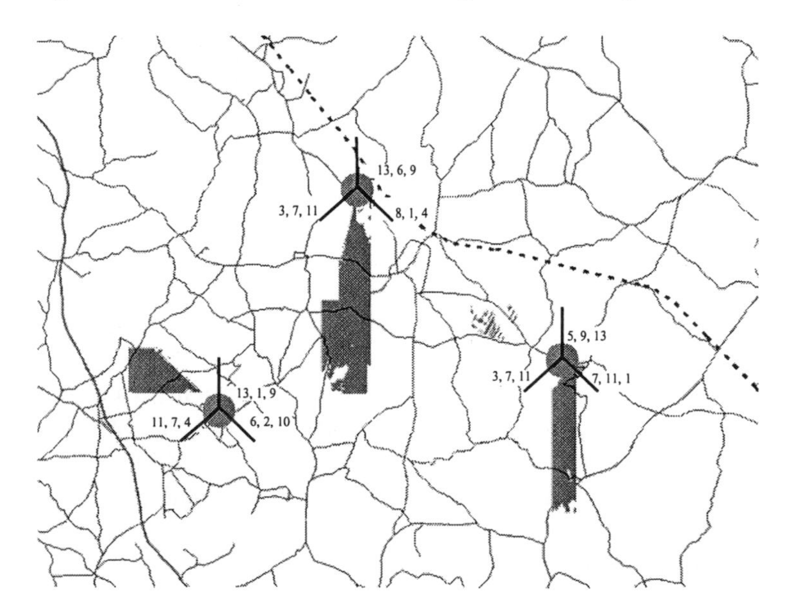

Figure 5.31. Too small a calculation region for interference analysis.

The impact of too small a calculation region to frequency planning and interference analysis is presented in Figure 5.31. Here the calculation regions are the same as in Figure 5.30. The frequency allocation is done by using the defined calculation areas. As can been seen there is not much interference, however the interfered areas are cut, so that there might be problems outside of the areas.

In Figure 5.32 calculation regions are defined larger but not large enough. Still, the interference situation is much different compared to Figure 5.30. On the right the interfered areas are cut which means that calculation areas are too small.

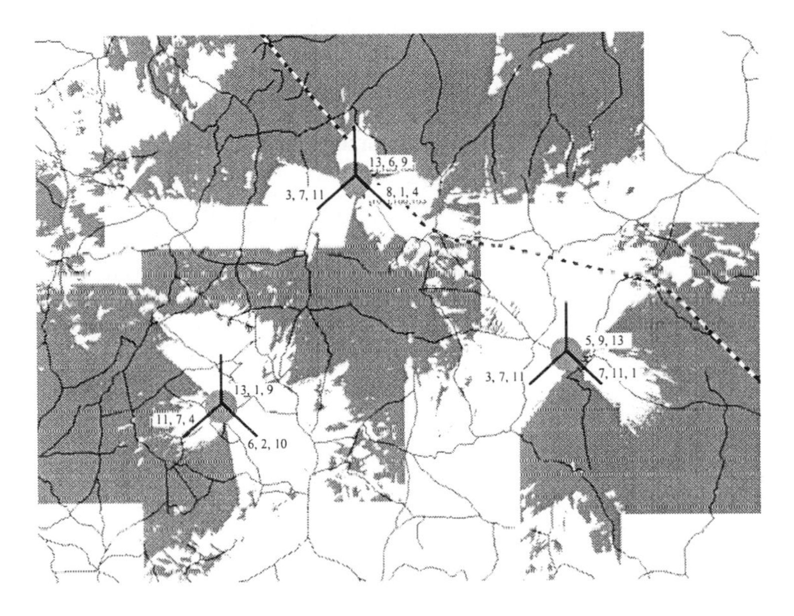

Figure 5.32. The interference analysis with larger calculation regions.

Figure 5.32 shows that it is very easy to get wrong results in the planning system if the basic definitions are not correct. The meanings of the parameters and definitions should be known before starting the planning. In this way major errors can be avoided.

5.7 Conclusions

See the summary of subjects and their outcome in Table 5.5.

Table 5.5 Radio propagation prediction.

Subject	**Findings**
Slope, decay index	–In urban areas slope is between 30–45 dB/dec, in rural areas closer to 30 dB/dec.
Okumura–Hata	–Commonly used propagation model. –Can be used for large cells, be careful in urban areas.
Juul–Nyholm	–Very similar to Okumura–Hata.
Effective antenna height	–The definition of antenna height depends on the topographics of the terrain. –Effective antenna height presents antenna height relative to the surroundings.
Model tuning	–The target is to find parameters for the basic propagation model and terrain type correction factors in a such way that the predicted field strength differs as little as possible from the measured field strength.
Model tuning measurements	–In model tuning measurements it is important that all the data is correct.
Digital maps	–Digital maps can have several layers depending on the information presented. Note that accuracy is not necessary the same as resolution.
Walfish–Ikegami	–Not a real micro cell model. Good for small macro cell.
Micro cell models	–Currently the micro cell models require a lot of computation time if the predictions are targeted to be accurate.
Antenna selection	–The target area affects the selection of antenna. Both horizontal and vertical 3 dB-beamwidths have an effect on the selection.
Calculation area	–Calculation areas for cells should be selected very carefully to achieve optimum results. Too large areas increase the computation time but too small areas give unaccurate results.

5.8 References

[1] Lee, W.C.Y., Mobile Communications Design Fundamentals, Howard W. Sams & Co., 1986.

[2] R. Janaswamy, "Radio Propagation and Smart Antennas for Wireless Communications," Kluwer Academic Publisher, 2000.

[3] Deygout, J., "Multiple Knife Edge Diffraction of Microwaves," IEEE Trans. on Antennas and Propagation, AP-14, no. 4, 1966, pp. 480-489.

[4] Hata, M., "Empirical Formula for Propagation Loss in Land Mobile Radio Service," IEEE Trans. on Vehicular Technology, vol. VT-29, no. 3, August 1980, pp. 317-325.

[5] COST-207, Digital Land Mobile Radio Communications, Annex 1, Final Report of the WG1 on Propagation, 1989.

[6] Recommendations and Reports of the CCIR, 1986, vol. V.

[7] Rathgeber, T., and Landstorfer F.M., "Land Cover Consideration in Mobile Radio Field Strength Prediction," 21st European Microwave Conference, 1991, pp. 1211–1216.

[8] Okumura, Y., Ohmori, E., Kawano, T. and Fukuda, K. "Field Strength and its variability in VHF and UHF land mobile radio service," Review of the Electrical Communication Laboratory vol. 16, nrs. 9–10, Sep–Oct, 1968, pp. 825–873.

[9] COST-231, Urban Transmission Loss Models for Mobile Radio in the 900 and 1800 MHz bands, TD(90) 199 Rev. 1, 1991.

Chapter 6

CAPACITY PLANNING

6. CAPACITY PLANNING

Capacity planning is typically done for the first time during the dimensioning phase and a second time in parallel with the coverage planning. The aim of capacity planning is to define the maximum and required number of transceivers at each base station or moreover the maximum and required capacity of the base station or base station site at each physical location over a certain area. The exact capacity need can be defined based on the monitoring of the radio network which gives the real information about traffic in the radio network. Thus the result of capacity planning is actually a specification for the frequency planning: this specification defines the required input information—the number of transceivers and the required base station site configurations to achieve capacity—for frequency planning at an element level.

Capacity planning is strongly related to the coverage planning process due to the coverage and capacity related base station site equipment like base station antenna line configuration (antenna height, beamwidth, tilting). Capacity and coverage planning are also linked to each other due to the propagation environment because frequencies can be reused more efficiently in a microcellular rather than in a macrocellular environment. Moreover, coverage planning defines the number of base stations required to cover a certain area and this group of base stations also defines the maximum radio interface capacity for this area when the maximum number of transceivers at each base station is specified. This definition of the maximum number of transceivers derives from capacity planning and it is based on the *frequency reuse number* or *factor*, meaning the number of base stations before the same frequency can be reused. (Note that *frequency reuse* means only how often the frequency is reused.) When the frequency reuse number is squeezed in an attempt to add transceivers to the base stations the aim is to avoid new base stations. If it is not possible to reuse the frequencies any more then new base stations are needed for the capacity. Thus, radio interface capacity can be extended by trying to add more transceivers by improving frequency planning or by adding more base stations. Assuming that more base stations are not required because of coverage planning, it is thus the most cost-efficient way to freeze the number of the base stations. *The target is therefore to keep the average antenna height as high as possible and simultaneously to calculate whether the number of the base stations is enough for the capacity if the number of transceivers is maximised.*

The key parameters for coverage, capacity and frequency planning are the *total traffic over a planning area* and the *average antenna height*. The aim is to divide the traffic homogeneously to all base stations inside the planning area and thus to use an equal number of transceivers at each base station. Capacity planning starts by specifying the target traffic (for example from monitoring results) and the minimum average base station antenna height to cover the area. The other essential parameter is the number of available frequencies (operator's frequency band) which defines the maximum number of transceivers at each base station.

An example is to be used to highlight the different phases of capacity planning. These phases are repeated in the different phases of the radio network evolution path: see Figure 1.14 that showed first that transceivers are added and then new base stations are implemented and then this process continues time after time. In radio network evolution it is good to remember that radio system planning starts from the macrocellular layer that should be retained as long as possible and only at the end of the evolution the small/micro base stations are profitable to deploy.

6.1 Capacity planning over a certain area

In the example, which is pursued throughout this chapter, there is a very high traffic density area of 5 km^2 in the urban area and four base station sites (BTS1 and BTS2 having three sectors and BTS3 and BTS4 having two sectors) of antenna height 25 m are needed to achieve good enough coverage for this area, see Figure 6.1.

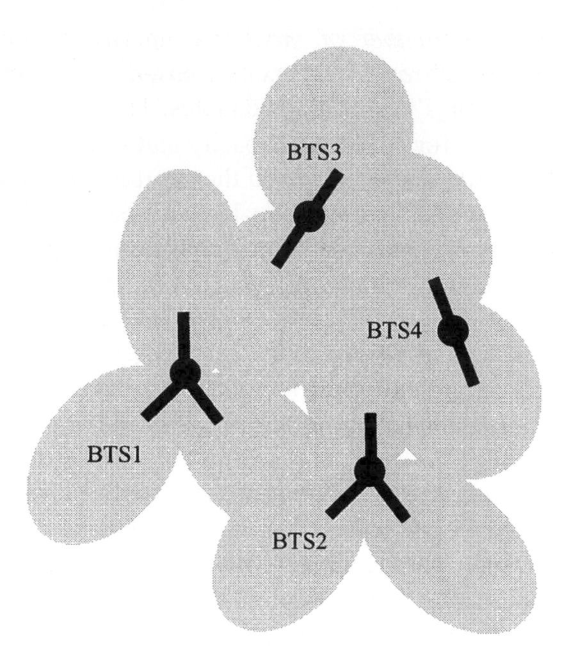

Figure 6.1. Base station sites in the capacity planning example.

The antennas are implemented at rooftop level and thus the propagation environment is of a macrocellular type. The traffic need for this area is as high as 100 Erl per busy hour. The frequency band is 6.0 MHz and this means 30 frequency channels in the GSM when the channel bandwidth is 200 kHz. The maximum number of transceivers at each base station depends on the frequency reuse factor that moreover depends on the propagation environment and the software features. The value 15 (frequency reuse factor) can be used in this example because it is a typical value for the radio network where antennas are implemented at rooftop level or above and when there are no special software features implemented. When the number of frequency channels and the frequency reuse factor are known, the maximum number of transceivers at each base station can be calculated:

Maximum number of transceivers
= frequency channels / frequency reuse factor
= 30 / 15
= 2.

These two transceivers represent a certain maximum traffic that has to be calculated in order to be able to define the maximum traffic that can be provided by the ten base stations (each having two transceivers). The calculation of the offered traffic by the two transceivers can be done by using Erlang-B or Erlang-C formulas and tables.[1]

6.2 Traffic models

Erlang-B (without queuing) or Erlang-C (with queuing) formulas and tables are based on a traffic theory [2] and the target is to estimate whether there is a channel available for the transmission (circuit switched speech or data). One transmission channel represents a unit of 1 Erlang if it is used over a hour and this hour is called a *busy hour*. When there are e.g. seven transmission channels in a common pool, seven transmissions can be made simultaneously without *blocking* and the eighth channel request is congested because of the lack of channels. Moreover, if these seven transmissions or calls take only 90 seconds each (this is an average value or a little bit higher for a speech call in GSM) the next new seven calls can be transmitted after 90 seconds. Thus, over a busy hour (= 1 Erlang) 400 (= 1 Erl / 25 mErl) calls can be transmitted per channel if all these calls happen exactly in a queue just after each other. However, in real life mobile users call randomly but while corresponding to the average traffic load 25 mErl. This random calling in time axis may cause blocking (if more than seven users are calling at the same time the eighth call is congested) and Erlang tables link a blocking and the maximum number of calls over the busy hour. Thus the maximum traffic or capacity at the base station has to be mentioned in Erlangs together with the blocking like 8 Erl with 1 percent blocking: the maximum traffic is also increased if higher blocking is allowed. Real speech traffic in radio networks follows the values of the Erlang tables extremely well but it still has to be remembered that these Erlang tables are answers only for the circuit switched traffic and when packet transfer data traffic comes in GSM (GPRS/EDGE/UMTS) new traffic models are required. The other thing to be remembered is that Erlang formulas and tables only give an estimate about radio interface capacity—the final results come from monitoring—but they are still good tools for capacity planning.

The Erlang formulas and ready-made tables directly tell of the traffic that can be offered by the different base station configurations. The Erlang traffic formulas have already been used in fixed mobile communications and their results can be used with confidence. The Erlang-B formula and table (the table is only one way to present the results of the formula) do not take into account the queuing that is typically used in the radio interface. However, the Erlang-B table is typically used in radio interface planning in GSM and queuing is reserved for extreme high traffic peaks.

Table 6.1. Erlang-B table for 1 percent and 2 percent blocking.

# of Channels	1 %	2 %	# of Channels	1 %	2 %
1	0,01	0,02	**26**	16,96	18,38
2	0,15	0,22	**27**	17,80	19,27
3	0,46	0,60	**28**	18,64	20,15
4	0,87	1,09	**29**	19,49	21,04
5	1,36	1,66	**30**	20,34	21,93
6	1,91	2,28	**31**	21,19	22,83
7	2,50	2,94	**32**	22,05	23,73
8	3,13	3,63	**33**	22,91	24,63
9	3,78	4,35	**34**	23,77	25,53
10	4,46	5,08	**35**	24,64	26,44
11	5,16	5,84	**36**	25,51	27,34
12	5,88	6,62	**37**	26,38	28,25
13	6,61	7,40	**38**	27,25	29,17
14	7,35	8,20	**39**	28,13	30,08
15	8,11	9,01	**40**	29,01	31,00
16	8,88	9,83	**41**	29,89	31,92
17	9,65	10,66	**42**	30,77	32,84
18	10,44	11,49	**43**	31,66	33,76
19	11,23	12,33	**44**	32,54	34,68
20	12,03	13,18	**45**	33,43	35,61
21	12,84	14,04	**46**	34,32	36,53
22	13,65	14,90	**47**	35,21	37,46
23	14,47	15,76	**48**	36,11	38,39
24	15,30	16,63	**49**	37,00	39,32
25	16,12	17,51	**50**	37,90	40,26

# of Channels	1 %	2 %	# of Channels	1 %	2 %
51	38,80	41,19	**76**	61,65	64,86
52	39,70	42,12	**77**	62,58	65,81
53	40,60	43,06	**78**	63,51	66,77
54	41,51	44,00	**79**	64,43	67,73
55	42,41	44,94	**80**	65,36	68,69
56	43,32	45,88	**81**	66,29	69,65
57	44,22	46,82	**82**	67,22	70,61
58	45,13	47,76	**83**	68,15	71,57
59	46,04	48,70	**84**	69,08	72,53
60	46,95	49,64	**85**	70,02	73,49
61	47,86	50,59	**86**	70,95	74,45
62	48,77	51,53	**87**	71,88	75,41
63	49,69	52,48	**88**	72,82	76,38
64	50,60	53,43	**89**	73,75	77,34
65	51,52	54,38	**90**	74,68	78,31
66	52,44	55,33	**91**	75,62	79,27
67	53,35	56,28	**92**	76,56	80,24
68	54,27	57,23	**93**	77,49	81,20
69	55,19	58,18	**94**	78,43	82,17
70	56,11	59,13	**95**	79,37	83,13
71	57,03	60,08	**96**	80,31	84,10
72	57,96	61,04	**97**	81,24	85,07
73	58,88	61,99	**98**	82,18	86,04
74	59,80	62,94	**99**	83,12	87,00
75	60,73	63,90	**100**	84,06	87,97

Part of the Erlang-B table is shown in Table 6.1 by concentrating on the blocking or so-called Grade-of-Service (GOS) values 1 percent and 2 percent that are generally accepted as a planning criteria for blocking in GSM.

Next, transceiver capacity (that is, the maximum traffic that the transceiver can offer) is calculated when the number of transmission channels also called *traffic channels* that can be used simultaneously for different mobiles or connections is defined. The radio interface of the GSM is based on *time division multiple access* (TDMA) which means that there are eight time slots at each frequency as shown in Figure 6.2.

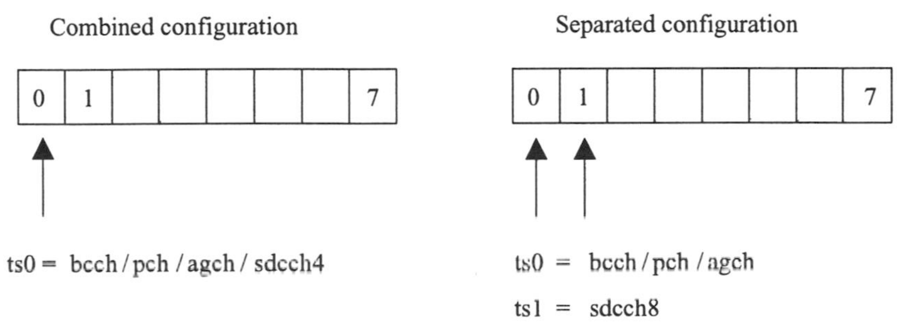

Figure 6.2. Time slots, traffic channels (TCH) and signalling channels (SCH) of the 1st transceiver.

These eight time slots can be used either for traffic (speech or data) or for signalling between the mobile and base station. Enough time slots have to be reserved for the signalling and the rest of the time slots can be used for traffic. Signalling requires one time slot (ts0) to send e.g. system information messages on the BCCH and paging on the PCH logical subchannels. This ts0 can also be configured to have 4 SDCCH (the combined configuration in Figure 6.2) which are needed for call establishment, location update and to send short messages. If more SDCCHs are required extra time slots (see the separated configuration in Figure 6.2) are needed to reserve enough SDCCH capacity. The need of SDCCH and moreover the need of signalling channels depends on the number of transceivers: typical planning criteria for the signalling channels is presented in Table 6.2.

Table 6.2. The number of the traffic and signalling channels in the different configurations.

TRX	TCH	SCH
1	7	1
2	15	1
3	22	2
4	30	2
5	37	3
6	45	3
7	52	4
8	60	4
9	67	5
10	75	5

In Table 6.2 it is assumed that only one time slot is required for signalling at a base station with two transceivers. This configuration used to function well when short messages were not used as much as today. However, the volume of the short message service today needs two signalling channels for base stations with two transceivers. Thus, there are 14 available time slots for the traffic (speech or data), the total traffic 7.35 Erl (at 1 percent blocking) of these 14 time slots can be observed in Table 6.1. The Erlang-B formula and table assumes that the traffic distribution is homogenous over a busy hour and thus does not work in situations where there are many simultaneous call attempts (the maximum number of simultaneous calls is now 14), for example after a concert, train-delay announcement or other occasion.

6.3 Frequency reuse factor for the macro base stations

The capacity of the individual base station was defined in the previous chapter and the total capacity of the ten base stations for the planning area can be extracted by multiplying 7.35 * 10 = 73.5 Erl. This value corresponds to the situation that all the base stations are equally loaded and there is a maximum blocking of 1 percent during a busy hour. It can be noted that this 73.5 Erl is not enough because 100 Erl was required and some spare capacity should also be available for the future. In order to expand the capacity more base stations would be required or more transceivers at each base station should be implemented. The implementation of new base stations is the most expensive solution and it should be avoided and thus one should concentrate on increasing the number of transceivers at the base stations (*to reuse frequencies more often or to squeeze the frequency reuse factor*).

6.3.1 Frequency reuse factor

The value of the frequency reuse factor depends on

- the *propagation environment* (antenna height versus rooftop levels = macro/micro level radio network, morphography and topography i.e. propagation slope)
- the *implemented radio network configuration* (especially on the antenna height and type distribution)
- the *software features* implemented in the radio network.

All these elements have a significant effect on the frequency reuse factor and the minimum frequency reuse factor can be obtained when these facts are known. The antenna height distribution has two effects on the frequency reuse factor: the lower the antenna height the smaller frequency reuse factor (due to *the propagation environment*) and the smaller deviation in the antenna heights the smaller frequency reuse factor (due to *the implemented radio network*). In the first case the man-made obstacles like buildings prevent radio propagation and this squeezes the frequency reuse factor values. In a macrocellular environment (base station antennas are clearly above the average rooftop level) buildings do not prevent the radio propagation too much and thus the minimum frequency reuse factor in the homogenous environment (that is, a totally flat area where the radio propagation properties are equivalent to all directions) is about 12. This means that the different frequencies have to be used at the 12 base stations and these frequency clusters (also called *frequency reuse pattern*) can then be reused, see Figure 6.3. In microcellular environments frequencies can be reused after 5–7 base stations not depending on the antenna heights, topography or morphography classes (only antennas have to be clearly below the rooftop levels).

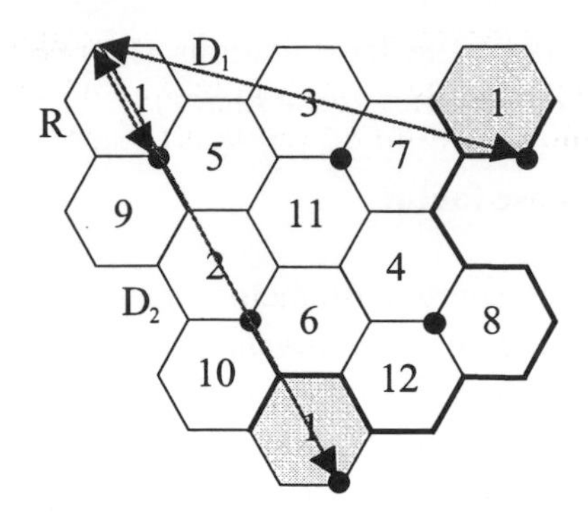

Figure 6.3. Frequency reuse cluster or pattern of 12 frequencies/frequency groups and two interferers (frequency one, highlighted).

In macro cells the environmental classes—topography and morphography—also have an effect on radio wave propagation and furthermore on the frequency reuse factor.[3] The radio propagation environment is not homogenous (there are higher and lower altitude positions, building density areas and several different area types like building types, parks, water areas) and thus the propagation environment is not equivalent in all directions. These discontinuous environment types may have better or worse propagation characteristics and thus the propagation slope varies and can increase the frequency reuse factor (or reduce the frequency reuse).

6.3.2 Frequency reuse factor and co-channel carrier-to-interference ratio (C/I_c)

The frequency reuse factor value 12 (in Figure 6.3) and 15 (in the example) were introduced for the macrocellular environment. These frequency reuse factor values are related to the co-channel carrier-to-interference ratio (C/I_c) and the reuse factor can be estimated when the C/I_c value is known. The reuse numbers and C/I_c values can be connected by calculating the received field strength levels from the different base stations and by analysing the required number of base stations before reusing the same frequency. This analysis can easily be done for the macrocellular propagation environment where radio propagation follows well the exponential attenuation $r^{-slope/10}$ (*r* equals the radius or distance from the base station and the slope equals the environment type) and where it is assumed that the base stations have quite constant coverage areas as depicted in Figure 6.3.

Each base station's (server and interferer) coverage area and the radius of the coverage area in the model in Figure 6.3 can be estimated by using path loss exponent and C/I_c from Equation 6.1 [1,3]

$$\frac{C}{I} = \frac{R^{-slope/10}}{\sum_n D_n^{-slope/10}} \qquad \textit{Equation 6.1}$$

Equation 6.1 shows that all interferers have to be taken into account in order to be able to calculate the final C/I_c. It also can be seen that Equation 6.1 is only a function of a slope and the horizontal and vertical (tilting) beamwidth of the base station antenna is naturally needed for the calculation in order to know the actual transmitted power. More accurate statistical C/I_c analysis requires including fast and slow fading in different environments.[4]

Tables 6.3 and 6.4,[3] show the calculated C/I_c values of the frequency patterns 9 (=9 frequencies) and 12 (=12 frequencies) when 120° and 65° horizontal beamwidth base station antennas without tilting are used and when the propagation environment varies from the urban to the rural (the path slope exponent varies between 2–4.5).

Table 6.3. The C/I_c values of the frequency reuse pattern 9.[3]

Slope 4	**C/I_c (dB)**			**C/I_c (dB)**	
Interferers	**120**	**65**	**Slope**	**120**	**65**
I1	19.2	20.3	2	4.2	6.2
I1, I2	18.8	19.8	2.5	7.2	9.3
I1, I2, I3	16.1	18.6	**3**	**10.0**	**12.3**
I1, I2, I3, I4	15.8	18.5	3.5	12.8	15.3
I1, I2, I3, I4, I5	15.6	18.2	4	15.6	18.1
I1, I2, I3, I4, I5, I6	15.6	18.1	4.5	18.2	20.9

Table 6.4. The C/I_c values of the frequency reuse pattern 12.[3]

Slope 4	C/I_c (dB)			C/I_c (dB)	
Interferers	**120**	**65**	**Slope**	**120**	**65**
I1	24.1	24.1	2	5.6	7.7
I1, I2	19.4	21.6	2.5	8.9	11.0
I1, I2, I3	18.9	21.1	**3**	**12.1**	**14.3**
I1, I2, I3, I4	18.8	20.9	3.5	15.3	17.5
I1, I2, I3, I4, I5	18.6	20.7	4	18.4	20.7
I1, I2, I3, I4, I5, I6	18.4	20.7	4.5	21.4	23.8

The first parts of Table 6.3 and Table 6.4 represent the influence of the number of the interferers on the C/I_c when the path slope exponent is constant and equals 4. The second parts of Tables 6.3 and 6.4 represent the influence of the path slope exponent (the number of the interferers is constant at 6). It has to be noted that the number of interferers has a great effect on the C/I_c values as can be seen in Tables 6.3 and 6.4. Moreover, the base station antenna type and the propagation slope influence significantly the C/I_c and thus also the frequency reuse number.

Table 6.3 and Table 6.4 show that a base station antenna of 65° in a horizontal plane gives better C/I_c values than a 120° antenna and thus a better (lower) frequency reuse factor can be obtained. Additionally it can be noted that C/I_c of 12.1 dB and 14.3 dB can be achieved when the reuse patterns 9 and 12 are utilised and when six interferers are included in the propagation environment where the path slope exponent is 3 (as in typical urban). These values exceed the threshold of 12 dB that is a criterion in the GSM specifications to avoid a co-channel interference in the non-hopping radio networks. The criteria for the co-channel and adjacent channel interference in the GSM are gathered in the Table 6.5.

Table 6.5. C/I_c and C/I_a criteria for the GSM radio networks.[5]

	Non-hopping	Hopping
C/I_c	12 dB	9 dB
C/I_a	–6 dB	–9 dB

In the capacity planning example the frequency reuse factor value 15 was selected and thus the C/I_c value should be even higher than 14.3 dB (Table 6.4, path slope exponent 3) if the 65° base station antennas are used. The value 14.3 dB represents a homogenous propagation environment where the exponent of the propagation slope is 3.0 and constant to all propagation directions without fading. The average value of

3.0 refers fairly well to the urban or suburban environment but this value is, of course, not constant in the real environment but varies and have a certain distribution. The selection of 15 was made because there has to be an additional margin ([14.3 dB – 12 dB] + 3 more frequencies) to partly take into account this inhomogeneous propagation environment and fast and slow fading. The correct margin for the different propagation environments has to be based on measurements; this value of the frequency reuse factor 15 is based on these measurements and the long-term experience of practical frequency planning results. The frequency reuse factor 15 gives an additional 3–4 dB margin for frequency planning but whether this is enough depends on the variations of the propagation environment. Additionally it is assumed in the model in Figure 6.4 and in Tables 6.3 and 6.4 that all the base station antennas are implemented at an equal height which is very theoretical. The variation in the base station antenna heights causes also strong variation in the propagation and thus it has a strong effect on the C/I_c and on frequency reuse in practise.

Next the frequency reuse number or factor should try to be squeezed in order to be able to add more transceivers to the example area and simultaneously to avoid implementing new base stations. This decreasing of the frequency reuse factor is called capacity planning or *capacity enhancement*. The alternatives to squeezing the frequency reuse number are

- to tilt the vertical beamwidth of the base station antennas
- to utilise software features or
- to drop the average base station antenna (macro→micro).

6.3.2.1 Tilting

Base station antenna tilting is the simplest way to limit the radio wave propagation in the vertical plane and this tilting, depicted in Figure 6.4, significantly increases the capability to reuse the frequencies (as the frequency reuse factor decreases). However, it has to be noted that too strong tilting also decreases the base station's coverage area. The amount of tilting (typically 5–15°) depends significantly on the base station antenna type. Sufficient tilting is 5–7° for high gain base station antennas (high gain antenna are equal to 15–18 dBi and vertical halfpower beamwidth <10° at 900 MHz) and around 10° tilting is required for the low gain base station antennas (low gain antenna are equal to 10–14 dBi and vertical halfpower beamwidth >>10° at 900 MHz). Tilting should be used in macro base stations almost without exception in the GSM in order to minimise interference and its meaning is still increasing in the UMTS. However, improvements in the radio network quality due to tilting have traditionally not directly be taken into account in capacity or frequency planning but has been used as an extra margin and to avoid serious interference areas due to inhomogeneous propagation in certain directions.

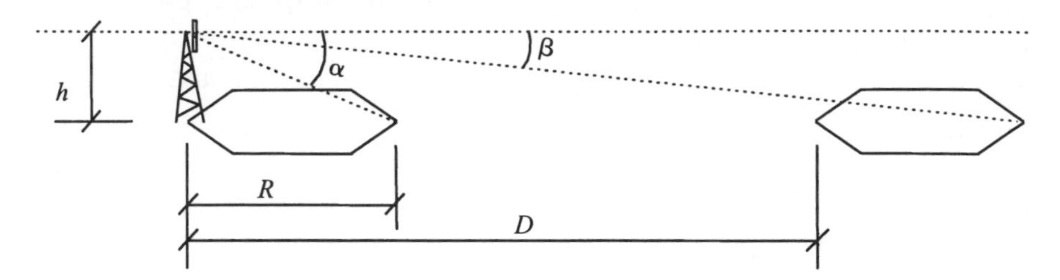

Figure 6.4. Base station antenna tilting.

6.3.2.2 Software features

The next possibility to squeeze the frequency reuse number is to utilise different software features for the radio network:

- *frequency hopping*
- *overlay–underlay radio network configurations*
- *interference cancellation* or *adaptive antenna* based solutions.

These features all aim to really squeeze the frequency reuse number such that a new lower reuse value can be used in capacity and frequency planning.

Frequency hopping is based on the usage of the several frequencies and thus on the interleaving of the transmitted information (speech or data) between these different frequencies. Frequency hopping should give at least 3 dB advantage (interference diversity) for the C/I_c as shown in Table 6.3 which refers to the values in the GSM specification. Thus, the frequency reuse factor could approximately be squeezed down to 12 which means a *twenty-five percent capacity increase*.

The second software related feature is an *overlay–underlay concept* that can be utilised in radio network implementation with or without frequency hopping. This overlay–underlay concept means that a part of the frequencies (called overlay frequencies) are used over the whole base station coverage area and a part of the frequencies (underlay frequencies) are only used close to the base station coverage area where there is a high field strength reception level (in other words a good co-channel carrier-to-interference ratio). It has to be noted that all the frequencies *can* be transmitted by using the same base station site configuration (base station and base station antenna line equipment) and they have exactly the same coverage area but the usage of them is limited to the certain areas based on the C/I_c values. Additionally, it has to be noted that different base station equipment suppliers have different solutions for this overlay–underlay concept but basically all these overlay–underlay concepts can be used in the macrocellular or microcellular propagation environment and thus the transceivers are not divided into macro or micro transceivers even if the name of this concept easily does.

The reduction of the frequency reuse factor in these overlay–underlay solutions is achieved by dividing the transceivers for overlay and underlay transceivers and by using the normal amount of frequencies for the overlay transceivers (e.g. a frequency reuse number of 15) and the smaller amount of frequencies for the underlay transceivers (e.g. a frequency reuse number of 5). The average frequency reuse number over the radio network is thus the weighted average of these 15 and 5 (the reuse numbers of the overlay and underlay). If there is 1 overlay transceiver and 1 underlay transceiver the average frequency reuse number is

$$(1 * 15 + 1 * 5) / 2 = 10.$$

This is even lower than in case of frequency hopping but it has to be noted that these reuse numbers for the overlay and underlay do not tell enough about the performance of this concept because the underlay can only be used when the C/I_c is good enough and thus the traffic distribution has a strong effect on the final performance.

The last software feature for macrocellular capacity enhancement is an *interference cancellation* or *adaptive antennas* in general. Interference cancellation is provided by the adaptive antennas and the target of this functionality is to improve the signal-to-noise (*S/N*) or carrier-to-interference (*C/I*) ratio by combining the received signal components of the individual antenna elements. Correspondingly, the frequency reuse number could *in theory* be decreased close to 1.0 if the fully adaptive antenna concept was used with beamsteering (*spatial division multiple access,* SDMA). Thus, the potential of the adaptive antennas for capacity increase in a macrocellular layer is enormous and clearly the highest among all the capacity enhancement software features. However, these adaptive antennas are not yet commercially available because they are complicated to manufacture and to adapt to the radio network.

6.4 Microcellular layer

Having introduced software features and their performances to give an idea of the capacity increase that these software features can provide for the macrocellular environment, it can be noted that the available techniques—frequency hopping, overlay–underlay concepts—may squeeze the frequency reuse number down to 10–12 and thus increase the capacity by 20–40 percent in a macrocellular environment. Some measurements may obtain even smaller values for the frequency reuse factor like 8–10 but it can still be agreed that the practical limitation for the reuse factor is around 10. In our example this reuse number 10 would mean 3 transceivers (= 24 time slots – 2 for signalling = 22 traffic time slots = 13.7 Erl at 1 percent blocking) at each base station and the total capacity of 10 base station would be

$$10 * 13.70 \text{ Erl} = 137 \text{ Erl}$$

that would already include 37 Erlang additional capacity for this service area. However, the traffic demand may exceed this maximum capacity and then there has to be a new solution to offer higher capacity. The last alternative is to build new base stations and step by step move from a macrocellular configuration to a microcellular configuration. This means new transceivers in new base stations but *simultaneously* it increases the radio network capacity because *the minimum frequency reuse number (5–7) is lower in a microcellular environment.* It has to be noted that the reuse factor of 5–7 corresponds to the capacity of 350 Erl in the example area. Moreover, the software features squeeze the reuse number in a microcellular environment in the same way as in a macrocellular environment. The microcellular radio propagation channel is still different from the macrocellular as explained in Chapter 1 and thus the performance of the software features is not necessarily equivalent in both environments. The adaptive antennas for example suffer in a microcellular environment with wide angular spread that complicates the filtering of unwanted signal components and this directly influences the performance.

6.5 Additional capacity related functions

There are also some individual functionalities in the radio network which may increase the capacity because they decrease interference. The power control and discontinuous transmission (DTX) are directly such features, and handovers could be classified indirectly as such a feature. Power control and DTX help greatly in the situations where there is major interference as in LOS connections in high buildings. It is clear that in these situations the interference level is decreased and it could also be said that capacity is increased. These improvements still happen only at certain locations and they cannot be said to be constant over the radio coverage area. Moreover, the influences of these features on capacity are difficult to measure and thus they cannot be taken into account in capacity planning but they could be utilised as an additional capacity reserve.

Handovers and handover parameters are also very important capacity related issues because the radio network capacity has to be fitted to the capacity demand on the field. This means that the base station's service (dominance) areas have to be well defined based on the traffic demand on each coverage area. This is called traffic distribution and will be discussed and explained in detail in Chapter 8 (optimisation) by concerning the even traffic distribution between the base stations.

Handovers also have an effect on the capacity in the dual band radio networks when the 1800 MHz layer is typically concerned as a capacity layer and the 900 MHz layer as a coverage layer. Thus, the target is to move all the dual band mobile users to the 1800 MHz layer and to use the 900 MHz layer only when there are no coverage in the 1800 MHz layer or when the 1800 MHz layer is not needed. This traffic distribution from the 900 MHz layer to the 1800 MHz layer happens easily by using C2 criterion in the GSM call set-up and handovers in preferring the 1800 MHz layer in the dedicated mode. Handovers are thus critical in a single frequency layer or dual band radio networks in order to utilise the available capacity but still they cannot be taken into account in the capacity planning.

6.6 Conclusions

Radio interface capacity planning and the performance of the main software features have been examined in a capacity plan for an example service area by squeezing the reuse factor and adding the third transceiver to every base station (whereby traffic was increased to 137 Erl). The future solutions for this area have been based on the micro base stations to increase capacity up to 350 Erl. The highlights of the capacity planning are gathered to Table 6.6.

Table 6.6. Capacity planning.

Subject	**Findings**
Traffic models	–Erlang-B for circuit switch traffic –Works well except in the mass call establishment
Frequency reuse factor	–Typical macrocellular reuse factor is 12 (theoretical) and 15 (practical) –Concentrate on the propagation and antenna heights
Tilting	–Reduce interference, use with high gain base station antennas –Use always
Frequency hopping	–Approximately 30–40 percent capacity increase
Overlay–underlay concepts	–Approximately 30–40 percent capacity increase
Adaptive antennas	–An extraordinary and complex concept –Reuse factor can be squeezed down to one
Additional capacity related features	–Power control and DTX are additional capacity reserve –Handovers and dualband parameters are critical to have good traffic distribution between the 900 and 1800 MHz layers

6.7 References

[1] W.C.Y. Lee, "Mobile Communications Design Fundamentals," John Wiley & Sons, 1993.

[2] W.C. Jakes, Jr., (ed.), "Microwave Mobile Communications," Wiley-Interscience, 1974.

[3] P. Seppälä, "Influence of the radio propagation channel on the frequency reuse factor in inhomogeneous environment," Masters Thesis, Helsinki University of Technology, 1998.

[4] R. Janaswamy, "Radio Propagation and Smart Antennas for Wireless Communications," Kluwer Academic Publishers, 2000.

[5] ETSI, Digital cellular telecommunications system (Phase 2+), Radio transmission and reception, GSM 05.05.

Chapter 7

FREQUENCY PLANNING

7. FREQUENCY PLANNING

7.1 Frequency planning criteria

Frequency planning is actually an implementation of capacity planning. Frequency planning together with capacity planning tries to maximise the information flow (voice or data) over the radio interface and simultaneously to maximise the efficiency of the radio network infrastructure. In cellular radio system planning the same frequencies are reused as often as possible in order to maximise capacity and thus minimise the radio network investments. The target is to have the maximum number of transceivers (a transmitter and receiver pair) at each base station without reducing radio quality. It has already been explained that frequency planning (together with capacity planning) begins with the specification of the required frequency channels (transceivers) at each base station. This work is related to the *frequency reuse factor* and was covered in detail in the capacity planning discussion.

When the basic radio network configuration—antenna heights, tilting, software features—is selected and optimised for the capacity demands, the minimum frequency reuse factor and the maximum number of the transceivers at the base station are defined and the frequency planning target has been defined. Next, this frequency planning target has to be achieved in frequency planning first by using the radio network configuration specified in the capacity planning (e.g. tilting, software features) and second by defining the frequency planning criteria.

Frequency planning criteria include the configuration and frequency allocation aspects. The configuration aspects consider the

- Frequency band splitting between the macrocellular and microcellular base stations
- Frequency band splitting between the BCCH and TCH layers
- Frequency band grouping
- Different frequency reuse factors for the BCCH and TCH layers.

All these configuration functions aim to ensure a good radio quality at least on a certain part of the radio network and to simplify the frequency planning process in order to avoid fatal errors as much as possible.

First the *frequency band can be split between the different base station types* based on the usage of these base stations. A certain frequency band can be reserved to be used only e.g. in the micro base stations and other frequencies on the macro base stations. This splitting is typically always mentioned when there are micro base stations in the radio network but there is not any good reason for this splitting in the frequency planning itself and the only reason that supports this splitting is the inaccurate coverage prediction in a microcellular environment. This coverage prediction in a microcellular radio propagation channel has been studied continuously and sufficient accurate prediction methods such as a ray-tracing prediction are under development (some products are of course already available). It still takes time before these ray-tracing predictions are implemented together with the automatic frequency planning algorithms to the same platform and before there is enough calculation power in the PCs. When coverage predictions are no longer an issue and they are accurate and quick enough, further frequency band splitting between macrocellular and microcellular layers is not necessary. Before this integrated macro and micro base stations' planning platform it is sensible to reserve a frequency band only for the microcellular layer especially if there are several micro base stations in the radio network. This splitting decreases the risks of a drop in radio quality crucially in the operative network due to the fatal errors of the micro base stations' coverage predictions in the planning tool.

Second a *frequency band splitting between the BCCH and TCH layers* (a layer means a group of frequencies which are used for a certain purpose e.g. only for the transceivers which include the BCCH time slot = BCCH frequencies) which is typically caused first by the impression that the BCCH layer requires more frequencies because of the BCCH time slot should be considered. The simulations show that this impression is not actually true because the information on the BCCH time slot is interleaved very widely and thus C/I_c values of 12–13 dB can be used on the BCCH time slot.[1] Moreover as has been shown, the TCH time slots on the BCCH frequency require C/I_c of 15–17 dB when frequency hopping is not utilised. Another reason for frequency splitting between the BCCH and TCH layers typically mentioned is to ensure a good radio quality on the BCCH frequency layer continuously and independently of the traffic load.

This is really an adequate reason to reserve a certain group of frequencies for only the BCCH layer and others for the TCH layer but there are still arguments for not splitting the frequency band at all. These are based on the idea of maximising the usage of all frequencies (reuse frequencies as often as possible) by providing a frequency plan as a result of the advanced and automatic frequency planning (AFP) algorithms and tools. These frequency planning tools provide, of course, considerably better plans than a human being because they have huge calculation capacities but still the risk might be too big to trust totally on the computer's intelligence. This risk due to traffic load and its control can be attempted on other TCH layers but not on the BCCH layer. Hence, it may be better in practise to reserve a separate frequency group for the BCCH layer and to use the rest of the frequencies freely on the other transceiver layers.

Next consider the usage of each single frequency as already studied; the maximum frequency reuse and the total frequency band divided into subbands. Each frequency can be used together with any other frequency from the other layer which gives the maximal flexibility for the frequency allocation but also adds the number of possible solutions and thus complicates the frequency allocation algorithms. The usage of the frequencies from the different layers can also be fixed by having certain frequency pairs or groups that can be used together. This *frequency grouping* is a restriction in a frequency allocation but simultaneously it may help to avoid too complicated calculations and to use e.g. adjacent frequencies at the adjacent base stations. Figure 7.1 shows a basic example of this kind of grouping when 15 frequencies are reserved for the BCCH, second TRX and third TRX layers each (frequency hopping is not used). When the grouping has been done there are a pool of 3 frequencies reserved for a certain base station when a frequency group is assigned in frequency planning. Thus, this grouping defines directly the frequencies that can be used at the same base station but it also minimises the adjacent channel frequencies at the adjacent base stations. An additional benefit is that all the frequencies are used evenly and the adjacent frequencies are not gathered over the same area, an advantage where frequency hopping performance depends partly on frequency separation. This frequency grouping however works well only if the radio network has been planned based on the coverage and capacity instructions explained thus far and by targeting to have an approximately equal number of transceivers (e.g. 2–3) at each base station. If there is 1 transceiver at some base stations and 5 or 6 transceivers at others this frequency grouping can not be used in the frequency planning but then the radio network configuration is already in a poor condition.

	f1	f2	f3	f4	f5	f6	f7	f8	f9	f10	f11	f12	f13	f14	f15
BCCH	1	2	3	4	5	6	7	8	9	10	11	12	13	14	15
2. TRX	15	16	17	18	19	20	21	22	23	24	25	26	27	28	30
3. TRX	29	30	31	32	33	34	35	36	37	38	39	40	41	42	45

Figure 7.1. Frequency reuse factor of 15 (BCCH, second TRX and third TRX) and frequency grouping.

The last item in the definition of the frequency planning criteria is the *uneven usage of the frequencies for the different layers*. More frequencies can be reserved for the BCCH to ensure at least one interference free frequency layer at each base station and less frequencies can be reserved for the other transceiver layers. The frequency grouping can also be done by using a different number of frequencies for the different layers as for example 15 for the BCCH, 14 for the second TRX and 13 for the third TRX. A microcellular layer (an overlay–underlay concept or separate micro base stations) can also be included and an example of this configuration can be seen in Figure 7.2. There are reserved 15 frequencies for the BCCH and second TRX layers and 6 frequencies for both microcellular layers. In this case the frequency groups have to be selected such that the adjacent micro layer frequencies are not allocated to the same base station site. If there are mainly 3-sector sites in the radio network the frequency groups of micro channels 31 (37), 33 (39) and 35 (41) have to be used in half of the sites and the frequency groups of micro channels 32 (38), 34 (40) and 36 (42) have to be used in the other half of the sites.

	f1	f2	f3	f4	f5	f6	f7	f8	f9	f10	f11	f12	f13	f14	f15
BCCH	1	2	3	4	5	6	7	8	9	10	11	12	13	14	15
2. TRX	16	17	18	19	20	21	22	23	24	25	26	27	28	29	30
1. Micro	31	32	33	34	35	36	31	32	33	34	35	36	31	32	33
2. Micro	37	38	39	40	41	42	37	38	39	40	41	42	37	38	39

Figure 7.2. Frequency grouping with the microcellular layer.

The first configuration in Figure 7.1 is meant for the radio network where baseband frequency hopping is utilized and the second configuration in Figure 7.2 is for the multilayer networks by utilising the overlay–underlay implementation strategy. With frequency grouping, frequency channel separations are considered automatically; when several frequencies are assigned to the same base station (typically the frequency separation is 400 to 600 kHz for 1–2 frequencies) or to the same base station site (adjacent channels are not assigned to the same base station site). If the frequency grouping is made intelligently and correctly the intermodulation can also be excluded by selecting the proper frequencies already in this grouping phase. Otherwise intermodulation has to be taken into account in the frequency allocation.

This intermodulation originates for example at the base station antennas because of the poor antenna element junctions which function as a mixer by producing intermodulation products e.g. 2f1–f2. The intermodulation source is worse in the downlink direction because the transmitting power levels are much higher than in the uplink direction. The intermodulation products are directed to the

- same base station's downlink frequencies (more than 2 frequencies at the same base station, $2f1_{down} - f2_{down} = f3_{down}$)
- same base station's uplink frequencies ($2f1_{down} - f2_{down} = f3_{up}$)
- adjacent base station's downlink frequencies (at the base station's border areas)
- adjacent base station's uplink frequencies (at the base station's border areas)
- 1800 MHz receiving frequency band from the 900 MHz transmission frequency band ($3f1_{down,900} - f2_{down,900} = f3_{up,1800}$) especially at the dual band antennas

and can cause deterioration in the C/I_c values and thus reduce radio quality. These intermodulations happen only when the transmission power is high and are worse when the transmission is reduced (power control) in the intermodulation product frequency. The GSM specification says that the co-channel ratio C/I_c has to be 9 dB or better in order to avoid interference. This means that if the intermodulation products are at the level of –30 dB and the C/I_c requirement is 9 dB the source frequencies can have 30 – 9 = 21 dB higher transmission power level. If the maximum power control is 30 dB in the GSM there is a real possibility of having an intermodulation interference at the same base station.

This intermodulation interference can be avoided by

- ensuring that the quality of the base station site equipment is high so that intermodulation does not exist

- grouping the frequencies such that the intermodulation products do not cause interference
- allocating the frequencies such that the intermodulation products do not cause interference

otherwise its complex influence on frequency planning can be made easier by

- preventing the power control
- directing the intermodulation products to the BCCH frequencies (there is no downlink power control on the BCCH).

Again it has to be noted that frequency grouping at the different frequency bands before the frequency allocation is the simplest, the fastest and maybe also the most cost efficient way of avoiding radio quality problems in the radio network caused by the intermodulation. Hence, the frequency grouping is maybe the most important planning criteria and also the most efficient tool for frequency planning.

7.2 Interference analysis and intereference matrix

The next step in the frequency planning process is to define the interference relations (co-channel C/I_c and adjacent channel C/I_a) between the different base stations at every locations in the radio network coverage area. These relations are the basis for the automatic frequency allocation and thus the definition of these relations have to be made accurately but simultaneously fast enough. The accuracy depends on the radio propagation and on the signal behaviour both of which can be calculated or measured. Calculation is preferred because measurement campaigns are typically very time consuming and sensitive to errors. Thus, interference relations are usually solved in the radio planning tool environment by utilising the radio propagation predictions and the result is the interference matrix where all the C/I_c and C/I_a connections can be found.

The essential fact for the success of frequency planning is the correct interference matrix while defining the correct interference matrix requires accurate coverage predictions and thus the propagation prediction models. This is the reason why prediction models should always be tuned because the radio coverage planning is worthless unless its quality is good. Accurate coverage predictions guarantee accurate frequency planning that furthermore guarantees good quality for the radio network. (The radio network can be planned without good coverage predictions and without good frequency plans and the radio quality may be reasonable but the radio network investment costs are typically much higher compared to the radio network with an accurate radio plan.)

Having discussed radio propagation predictions and emphasising the requirement for accuracy in order to produce accurate coverage predictions, the received signal or actually signals (from the different base stations in the downlink and from the different mobiles in the uplink directions) that have an effect on the *C/I* (co-channel or adjacent channel) predictions must be considered. In coverage planning there is only one serving signal at each mobile station location and the standard deviation of this signal has been measured and studied and it is well-known that a typical value is 7 dB at 900 MHz frequency band.[2,3] In the interference analysis (and in the real networks) there are multiple co-channel or adjacent channel signals received by the base station or mobile station and thus the combination of these received signals no longer has the same standard deviation as the single signal has.[2] Figure 7.3 shows the calculated mean and standard deviation (std) values for the *N* signals of 0, 6, 8.3 and 12 dB standard deviation. If the signal std is 6 dB and only one signal is received, the mean value of this received signal is 0 dB and the std of this signal is 6 dB, and correspondingly if there are two signals of 6 dB std, the mean increases to 4.58 dB and the std of the total reception is 4.58 dB.

	0 dB		**6 dB**		**8.3 dB**		**12 dB**	
N	**mean**	**std**	**mean**	**std**	**mean**	**std**	**mean**	**std**
1	0.00	0.00	0.00	6.00	0.00	8.30	0.00	12.00
2	3.00	0.00	4.58	4.58	5.61	6.49	7.45	9.58
3	4.50	0.00	6.90	3.93	8.45	5.62	11.20	8.40
4	6.00	0.00	8.43	3.54	10.29	5.08	13.62	7.66
5	7.00	0.00	9.57	3.26	11.64	4.70	15.37	7.13
6	7.50	0.00	10.48	3.04	12.69	4.41	16.74	6.74

Figure 7.3. The multiuser/multisignal interference.[2]

Note that these two results are quite different and thus the influence on the frequency allocation should be significant. In addition it should be noted that these values are valid only when the signals are at an equal level and this partly reduces the effect of this multiuser interference. However, there are plenty of locations in the radio network where two co-channel signals are interfering with the serving channel. The open questions are the total influence of the multiuser interference and whether these locations should be taken into account in the frequency allocation or not.

The other issues in the interference analysis and in the creation of the interference matrix are radio propagation and the base station relations. The main questions are the radio propagation attenuation (slope) and the calculation area for the interference analysis. Errors in the radio propagation slope may cause huge variations in the received power levels and thus deteriorate the *C/I* accuracy significantly.

The interference calculation area should be much larger than the base station dominance area (that is, the area where the base station's service is superior) or coverage area (the area where the base station provides service; the power level that exceeds the defined coverage planning threshold). Moreover, the calculation time increases enormously at the same time. In order to reduce the calculation time the interference analysis is typically done only in the area where the base station coverage calculation areas overlap. Thus, it is critical to define large enough calculation areas for all base stations (see Figure 7.4). Even if the power levels are quite low in the calculation border areas and it seems that these low power levels are not used, there may still be need to use these for service in indoor locations and there too interference has to be avoided.

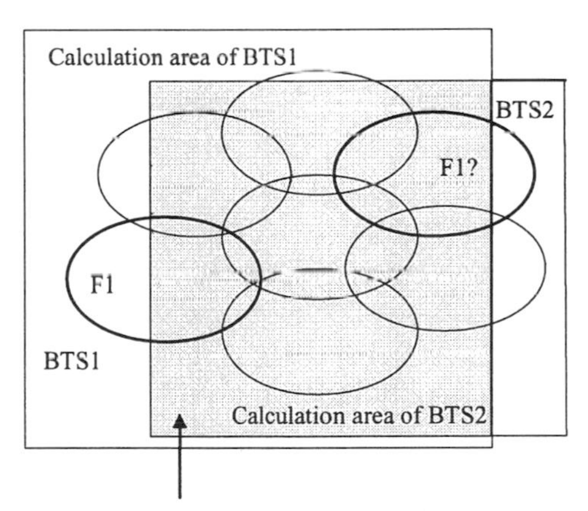

Figure 7.4. Interference calculation area.

An interference matrix could also be created by measuring base station relations from the radio network. This measurement method is very time consuming process and also sometimes inaccurate because interfering base stations may be very far away and connected to a base station controller other than the measured base stations. This measurement method is still applicable in urban and especially the microcellular environment where coverage predictions are not accurate and where interfering base stations are not too distant.

7.3 Automatic frequency planning (AFP)

Frequency planning and specifically automatic frequency planning (AFP) is required in order to maximise the frequency reuse and radio quality in the radio network. Frequency planning can be done manually without using any frequency planning tool but the result is rarely optimised because of the limited calculation capacity of a radio planning engineer. Automatic frequency planning tools can process complicated calculation algorithms and the result is based mainly on the performance of allocation algorithms and the computational capacity. The performance and risk of the frequency allocation can be improved and optimised by frequency grouping and thus directed to exclude intermodulation and to automatically reduce the adjacent channels at adjacent base stations.

Frequency planning is also a continuous process in that it has to be repeated time after time for different areas where as it is not worth making a new frequency plan for the whole radio network every time. The frequency plan is only optimised for a short period and subsequently starts to deteriorate, as presented in Figure 7.5. Typically a frequency plan upgrade has to be done for a certain area where there are new transceivers or new base stations launched. Every partial upgrade brings the frequency plan quality away from the optimised frequency plan, as shown in Figure 7.5. Moreover, it is expensive and very time-consuming to make a frequency plan upgrade every time for the whole radio network. Thus, the number of frequency plan upgrades should be minimised (all the changes or as many as possible should be made at the same time) and after a certain period e.g. 6–12 months a new frequency plan should be introduced. This new plan should concern at least the urban and suburban areas in order to achieve the performance level of the optimised frequency plan and thus to maximise frequency reuse in these areas.

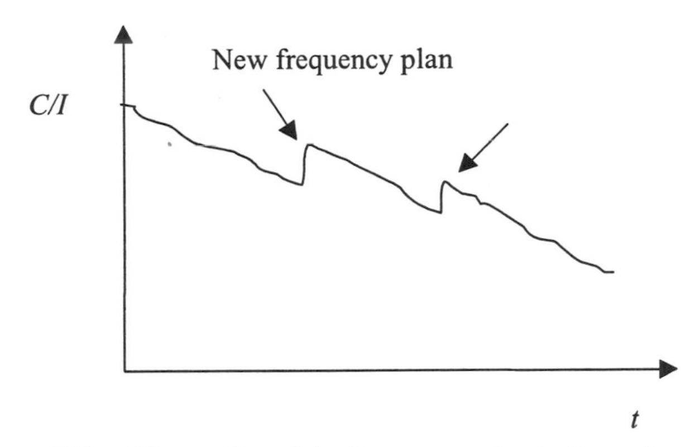

Figure 7.5. The quality of the frequency plan in the radio network.

The aim of frequency planning is to provide an equal interference level over the whole radio network in order to guarantee even radio quality for the subscribers. Critical for the equal interference level are the base station boundary areas and, therefore indoor locations.

Frequency planning or allocation begins with the definition of restrictions in the use of certain frequencies or frequency groups at the same base station or at the same base station site. The restriction in the use of frequencies at the same base station depends on the combiner (that combines all the frequencies at the base station to the same cable) which typically can not combine frequencies that have a separation less than 400–600 kHz (2–3 channels separation in the GSM). Thus, for example, channels 99 and 101 can not be used at the same base station with channel 100. The other restriction in the use of frequencies at the same base station site is related to co-channel and adjacent channel interference. It is not appropriate to use these co-channel or adjacent channel frequencies at the same base station site in order to avoid continuous interference. Thereby, if channel 100 is already assigned to the base station, channels 99 and 101 should be avoided at the same base station site (including all cells at this site). These frequency channel restrictions can mainly be avoided again if the frequency grouping is made before the frequency allocation (as mentioned before) when there are less restrictions to take into account by the frequency planning tool and thus the complexity and the risk of frequency allocation is effectively minimised.

Finally, the preliminary preparations have been completed and it is time to start the actual frequency allocation. The allocation can be made on one occasion if the frequency band is not split—for example for the

BCCH and TCH layers or macrocellular and microcellular layers. If the frequency band is split every layer requires its own allocation. Next confirm whether all frequencies are taken into account in the allocation or whether the frequency band is grouped and thus initially allocated. If this is the case the frequency allocation is better done based on the frequency groups (for example by using the BCCH frequency as an ID for each frequency group) in order to minimise the aforementioned complexity and risks in the frequency allocation.

After defining the frequencies to be included in the allocation based on frequency band splitting or frequency band grouping the last action is to define the target values for the frequency allocation and the weights for the uplink and downlink directions as well as for the different layers (in the case that all frequencies are used in the same frequency pool and not grouped). The target values (also called thresholds) are set for the co-channel and adjacent channel interference and typical values are for example 3 percent co-channel interference and 5 percent adjacent channel interference. It can also be noted that these co-channel and adjacent channel interference can be weighted e.g. by setting the co-channel interference threshold to 1 percent and the adjacent channel interference to 7 percent. Correspondingly, the uplink or downlink directions and also the different layers can be weighted which is especially important when all of the frequency channels are used in all the frequency layers.

7.4 Additional features in frequency planning

7.4.1 Frequency hopping

Frequency hopping improves the *C/I* and frequency reuse factor and thus adds capacity into the radio network. The final capacity improvement of frequency hopping depends on the number of channels and the frequency bandwidth. The implementation of frequency hopping also has a significant effect on frequency planning which furthermore depends strongly on the frequency hopping scheme. The *baseband* and *synthetized* frequency hopping schemes are typically utilised for example in the GSM mobile networks and frequency band limitation or sufficiency, as well as system limitations, have a significant effect on the selection of these frequency hopping schemes for different purposes.

Baseband hopping

Baseband hopping is an elementary feature of the GSM system and its usage is not typically limited. Baseband hopping means that the call (the radio connection between the mobile station and base station) is switched between different transceivers (or frequencies) periodically or randomly during the connection. This means that the call uses a different transceiver from the pool of all the transceivers at each time slot. If there are three transceivers (frequencies f1, f2 and f3) the call proceeds f1→f2→f3 →f1 ... on different time slots when *cyclic* hopping is used. Figure 7.6 specifies the baseband hopping sequences,[4] because note that the BCCH time slot can not be hopped as it has to be sent on the same frequency all the time. There are two different defined baseband hopping groups which are the group 1 including all the time slot zeros except the BCCH time slot and the group 2 which includes all time slots 2–7. If there are four transceivers/frequencies at the base station in total it means that three of them are used in the first group and all of them are used in the second group. The usage of these frequencies can be cyclic as in Figure 7.6 or pseudorandom.[4] *Pseudorandom hopping is preferred because of the SACCH blocks which are sent by an even period.*

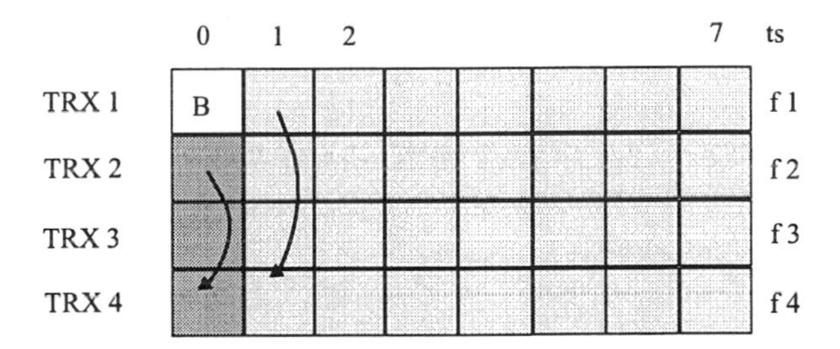

Figure 7.6. Frequency hopping groups in the baseband hopping.

Note that the number of frequencies to be utilised in baseband frequency hopping depends on the number of transceivers (number of frequencies in baseband hopping equals the number of transceivers). Because the performance of frequency hopping depends on the frequencies to be hopped it means that baseband hopping requires quite a wide frequency band before it is efficient. If the frequency reuse factor was 12 and three frequencies are required at each base station, 36 (7.2 MHz) frequencies are needed. Frequency diversity is reasonable when two frequencies are used for baseband hopping,[5] but the improvement of interference diversity is significant only when there are at least three frequencies at the base station.[4]

The requirement of the wide frequency band is the weakness of baseband hopping and thus it is typically not used in the radio networks of narrow frequency band. The advantages of baseband hopping are

- the usage of the BCCH frequency (except time slot 0)
- the improvement in the frequency reuse factor and thus in the capacity
- easy frequency planning.

Because baseband hopping can be utilised on the BCCH traffic time slots, the frequency reuse factor can easily be reduced down to 12 on the TCHs of the BCCH frequency and on other frequencies. When we recall that the C/I_c requirement (which is approximately the same as the frequency reuse factor) for good radio quality is around 12–13 dB on the BCCH time slot [1] it can be concluded that baseband hopping makes it possible to have the same C/I_c for the TCH time slots as for the BCCH time slot on the BCCH frequency. Thus, the same frequency reuse factor (for example 12) can be used as a planning criteria for the BCCH as for the other frequencies.

The frequency planning process with baseband frequency hopping is equivalent to the normal frequency planning process. The target frequency reuse factor has to be defined initially in order to know the maximum number of frequencies at each base station. Because frequency hopping reduces the frequency reuse factor as a function of the number of transceivers both the targeted number of transceivers and then the frequency reuse factor have to be decided. Next, the frequency allocation has to be done based on the targeted frequency reuse number which can be transferred to the C/I_c value which moreover has to be used as a target value in the frequency allocation.

Synthetized hopping

Synthetized frequency hopping differs from baseband hopping such in that there are both more system requirements and flexibility in synthetized hopping. First of all synthetized hopping means that the frequency can be changed in each transceiver between time slots. Moreover, this means that the whole frequency band can be reused in each transceiver which means that a wideband combiner is required to combine the frequencies at the base station. This wideband combiner typically causes the first limitation for synthetized hopping namely that the maximum number of transceivers is restricted. Second, the BCCH frequency can not be included in synthetized hopping (see the Figure 7.7) because the BCCH time slot has to be sent at the same frequency all of the time; the BCCH transceiver's frequency can not be changed.

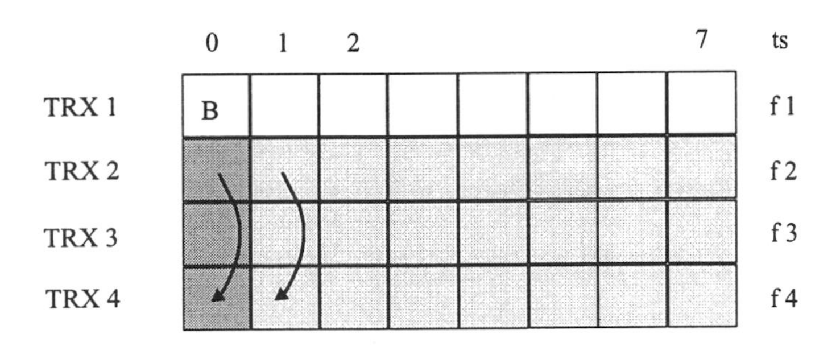

Figure 7.7. The frequency groups of the synthesised frequency hopping.

Even if the system and radio network elements limit synthetized hopping there are interesting flexibilities in this hopping scheme because all of the frequencies can be used in each transceiver which increases the performance of the synthetized hopping. This method, to use all the frequencies in all the base stations, is called a 1/1 pattern (frequency reuse factor), the other commonly mentioned pattern is the 1/3 which means that the frequency band is divided into three groups. These values describe only the reuse of the frequencies but the practical and minimised average frequency reuse factor is still around 7–9 for synthetized hopping before the radio quality drops (the frequencies are used at all base stations but not simultaneously and thus some of the transceivers can not be used all the time: fractional loading). This also means that the number of the transceivers at each base station has to be dimensioned based on the average frequency reuse factor in order to maximise the radio network efficiency. The performance of synthetized hopping is best with the narrow frequency band when it is not possible to site many transceivers at each base station and thus baseband hopping can not be used efficiently. Table 7.1 shows an example of the required frequency reuse factors if the same frequency band were to be used with baseband and synthetized hopping. It can be seen that the requirement of the reuse factor of 7 is very tight even for synthetized hopping.

Table 7.1. Typical frequency reuse factors for TRX layers.

	BCCH_TRX	other_TRX	Average
Baseband hopping	12–13	12	12
Synthetized hopping	15–17	**7**	**11–12**

Finally, baseband hopping specifically, but also synthetized hopping, influence frequency splitting. In the case of baseband hopping all of the frequencies are mixed and only the BCCH time slots use one separately defined frequency. Thus, frequency splitting works only for the BCCH time slot. Correspondingly, frequency splitting works for the whole BCCH frequency layer (BCCH and TCH time slots) in synthetized hopping but the other transceivers can not be easily separated.

7.4.2 Discontinuous transmission (DTX)

There are at least two features—discontinuous transmission and power control—in the radio interface which improve frequency efficiency and furthermore the radio quality but which are typically not utilized in frequency planning. Discontinuous transmission (DTX) can be used in the uplink direction in order to decrease interference. A voice detector (VD) in the mobile phone detects whether the user speaks or not and gives an indication to the mobile station's transmission part. When the user is not speaking the transmission part only sends data in a part of the time slot and thus the transmission is disconnected and interference is reduced. The voice detector recognises when speech restarts and simultaneously normal transmission occurs.

It is obvious that this kind of function reduces interference and thus improves the quality and frequency band efficiency. However, it is very difficult to measure the improvement of this discontinuous transmission, for example in decibel units, in order to be able to take DTX into account in frequency planning. Moreover, DTX evaluation measurements require a dense traffic radio network and the improvement has to be unambiguous. It is globally discussed that the DTX improvement is 3 dB based on simulations or measurements. This value could be utilised in frequency planning by reducing the required *C/I* value but it is not typically used; this 3 dB is specified as an additional margin to ensure radio network quality.

7.4.3 Power control

Power control is a similar function to discontinuous transmission when concerning frequency planning: it is obvious that it reduces interference but it is still not to be utilised in frequency planning because its effect on the planning criteria is unclear. The aim of power control is to minimise the base station's and mobile station's power consumption and thus also to minimise transmission power and interference. In addition, power control's aim is to have a sufficient good and constant field strength level at the base station's and mobile station's reception locations where the received field strength may be much higher. This means in other words that all the situations where too much power is used for maintaining the base station–mobile station connection are noted and this extra power is filtered.

The performance of power control and moreover its influence on the interference level in the radio network depends strongly on the power control strategy. Power control works by defining the maximum (upper limit) and minimum (lower limit) receiving values before starting to increase or decrease the power, see Figure 7.8. The aim is to have the received field strength level between these limits and power is increased or decreased by steps in order to keep the receiving level between these limits. If there is a huge difference between the limits, the transmission level can vary and it also takes time to adjust the optimised transmission level. In this case the interference reduction is also lower and only the highest interference peaks are filtered. Correspondingly, power control is very quick and efficient in reducing interference if the window between the limits is narrow. This strategy causes power control to happen often (loads the base station controller) and its influence on the handovers has to be taken into account.

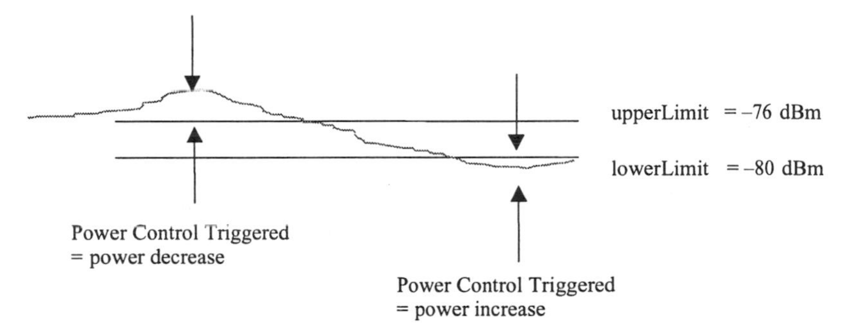

Figure 7.8. Upper and lower power control limits for very tight power controlling.

Even if the power control strategy is well-known it is still difficult to measure the power control's performance and even more difficult to define a power control planning criteria and thus it is better to have power control improvement as an additional margin for frequency planning. Moreover, power control is better understood as a feature to avoid the extreme interference areas in variable environments such as in urban areas.

7.4.4 Multilayer radio networks

Frequency band efficiency can be improved by introducing multilayer radio networks in order to increase capacity in radio interface. The multilayer radio network means that physically the base station coverage areas totally overlap as shown in the Figure 7.9 or frequencies (transceivers) are used only at certain locations at the base station coverage area. When the coverage areas overlap and different base stations are used, these base stations are named as macro or micro base stations based on coverage area size (depending mostly on the antenna height). When a certain frequency is used inhomogeneously at the base station's coverage area, this frequency is named as an overlay (can be used over the whole coverage area) or underlay (can be used only very close to the base station when the *C/I* is good enough, can also be called a micro layer) frequency.

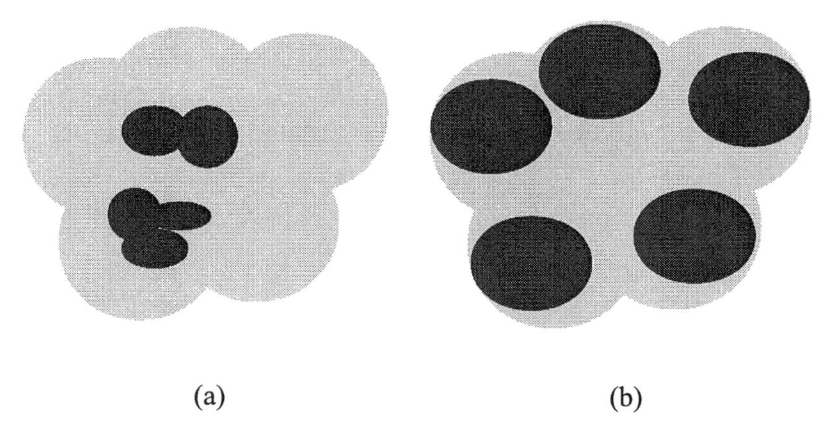

(a) (b)

Figure 7.9. Multilayer radio networks. (a) macro and micro base stations, and (b) overlay and underlay frequencies.

When micro base stations are introduced in the radio network, the frequency band is typically split and the frequency assignment for these micro base stations is done manually based on the measurements. This manual assignment has been used because propagation models are not good enough and they can not model the radio wave propagation. The micro base stations or frequency band splitting have no major influences on the frequency planning parameters or frequency planning process.

The overlay–underlay multilayer strategy has a stronger influence on frequency planning because frequency planning thresholds (*C/I* target values) have to be re-defined and frequency grouping has to be done by taking into account that overlay and underlay frequencies do not interfere with each other and also to group underlay frequencies at the same base station site. These micro transceivers (underlay frequencies) have been the subject of the frequency grouping exercise throughout this chapter. Having selected 6 frequencies for each underlay transceiver layer in order to avoid using the same frequencies at the same three base station sites, these frequency splitting and groupings of the normal frequency planning process can be carried out using the relevant *C/I* target values.

7.5 Conclusions

Having discussed frequency planning and the key frequency planning instructions in order to guarantee high capacity and high quality for the radio network, the meaning of the frequency planning criteria and especially the meaning of *frequency grouping* was emphasised, which offers a solution to avoid interference caused by the intermodulation. The highlights of this chapter are gathered in Table 7.2.

Table 7.2. Frequency planning.

Subject	**Findings**
Frequency reuse factor	–A result of the capacity planning
Frequency splitting	–Use only if required
Frequency grouping	–Helps e.g. in the co-channel interference, adjacent channel interference, intermodulation, frequency hopping
Intermodulation	–Cause easily a bad radio quality –Can be tackled by the frequency grouping
Interference matrix	–A key for a successful frequency allocation
Multiuser interference	–Cause uncertainty
Frequency allocation	–Evaluate the allocation tool –Good coverage predictions are needed
Baseband hopping	–Easy and beneficial to utilize
Synthetized hopping	–Complex—too many limitations
Discontinuous transmission (DTX)	–An additional margin –No influence on the frequency planning
Power control	–An additional margin –No influence on the frequency planning
Multilayer networks	–An alternative for the frequency hopping –Different frequency plans for different layers

7.6 References

[1] J. Wigard, T.T. Nielsen, P.H. Michaelsen, P. Mogensen, "BER and FER prediction of control and traffic channels for a GSM type of interface," IEEE Proceedings of VTC98, Ottawa, Canada, 1988.

[2] Petri Seppälä, "Influence of radio propagation channel on frequency reuse factor in inhomogeneous environment," Masters Thesis, Helsinki University of Technology, 1998.

[3] D. Parsons, "The Mobile Radio Propagation Channel," Pentech Press, 1992.

[4] J. Haataja, "Effect of Frequency Hopping on the Radio Link Quality of DCS1800/1900 System," Masters Thesis, Helsinki University of Technology, 1997.

[5] A. Nieminen, "Influence of Frequency Hopping on the Frequency Planning in Indoor Locations," BSc Thesis, Häme Polytechnic, 2000.

Chapter 8

OPTIMISATION

8. OPTIMISATION

The radio system planning process as described in Chapter 2 mentioned that optimisation and monitoring is the last part in radio network planning. Monitoring was linked to optimisation because they both partly include the same measurements. Monitoring results are needed and used during the whole radio planning process and thus optimisation is actually the real final phase in the radio network planning process. By the optimisation phase the coverage, capacity and frequency plans have already been carried out and implemented and the monitoring results can be measured from the operative network. Hence, the primary purpose of optimisation is to verify and improve the actual radio plan by taking into account the radio network evolution as

- subscriber growth
- radio interface traffic
- coverage demands (indoor/outdoor coverage)
- radio quality
- overall radio network functionality.

Optimisation considers all these individual topics in the different optimisation phases which can be presented in two main areas called *network assessment* and *network verification and tuning* as shown in Figure 8.1.

	Quality			**Functionality**	
	Coverage	Capacity	Interference	Handovers	Drop calls
A: Network Assessment	dBm	Erl	BER	failure %	%

	Configuration Management			**Functionality**	
	Coverage	Capacity	Interference	Handovers	Drop calls
B: Network Verification and Tuning	BTS	TRX	Frequency	Radio Parameters	Radio Parameters

Figure 8.1. Optimisation areas and phases.

It can be seen in Figure 8.1 that optimisation work mainly concerns the cost-efficiency and overall quality of the radio network, again. Optimisation is required because the subscriber requirements, base station functionalities and available frequency band are continuously changing. The radio network assessment phase in Figure 8.1 can also be called *pre-optimisation* or an *analysing phase*. In this analysis are gathered the key performance indicators and performed field measurements from the specific radio network area in order to find the optimisation areas (coverage, capacity, interference, functionality) where the radio network should be improved. The key performance indicators (see Chapter 9) monitor all the key areas in the radio network and they basically reveal if the cost-efficiency or quality (coverage, capacity, speech/data transmission quality; off-set against the interference level) are low in the radio network.

Next the different planning areas have to be verified in order to find detailed reasons for coverage, capacity, interference or functionality problems and this phase is called network verification and tuning. In this verification and tuning phase the radio network configuration is also adjusted to achieve radio system planning targets. The network tuning contains the final problem solving actions like

- to change the base station site configurations
- to tilt antennas, to turn antenna direction, to change antennas
- to change the radio parameters
- to build higher or lower antenna masts
- to move antenna locations
- to move base stations

in order to change the base station coverage or dominance areas to balance the traffic (to improve the usage) or to improve the quality or handovers and thus overall *call success rate*. The required actions have to be clarified step by step by trying always to avoid moving base stations or building new base stations which may in the worst case be the only possible solutions to improve the radio network quality.

8.1 Optimisation tools

In order to be able to perform successful optimisation work some essential optimisation tools are needed. These optimisation tools can be divided into three categories like

- database tools
- radio planning tools
- measurement tools.

Database tools are required to manage the base station site configurations. The essential radio planning parameters for (for example) coverage planning are the *antenna height, antenna type, antenna direction, antenna tilt, cable length and type, diversity technique, LNA parameters, combiner type* and *BTS transmission RF peak power.* These parameters can not automatically be found from the same database and there is considerable work ahead in the optimisation phase if these base station site configurations have not been managed properly.

Radio planning tools are, of course, needed to calculate the base station coverage or dominance areas and especially for the coverage plan post-processing to predict interference and to allocate frequencies to the base station sites. In some cases the advanced radio network simulators could also be useful if they work properly.

Measurement tools are required to analyse the base station traffic, coverage, speech quality and handovers to the neighbour base stations in the *downlink and uplink directions*. There are several tools on the market available to fulfil these requirements and a set of these tools have to be selected based on the available functions. Note that different tools are typically required for the downlink and uplink measurements. Network Management System (NMS) typically gathers data in the uplink direction and the mobile/PC combination is used to measure the downlink direction.

The NMS typically measures the capacity (traffic in Erlangs and blocking) and the function of the radio network (handovers, handover failures and drop calls) over a certain area at any time of the day on a base station basis or area basis. Moreover, coverage and quality can be measured together in the downlink direction (NMS measures in the uplink direction) and these measurements have to show

- field strength
- bit-error-rate (BER) in the radio networks without frequency hopping
- frame-erasure-rate (FER) in the radio networks with frequency hopping
- frequency channel
- frequency channels of neighbour base stations
- measurement location.

These measurements are first used in the network assessment phase and they are essential when studying the radio network Quality of Service (coverage, capacity, interference and functionality) in detail.

8.2 Radio network assessment

In a radio network assessment field measurements of coverage, capacity and interference can usually be measured simultaneously. First selected is an optimisation area (for example based on the key performance indicators) which contains a group of base stations (around 10–20). Next, the base station (coverage or) dominance (serving area) areas are measured and simultaneously interference (bit-error-rate (BER) or frame-erasure-rate (FER)) and call and handover success rates which indicate potential radio channel blocking or handover problems. All these measurements are done in the downlink direction by using the mobile/PC combination to determine the radio network performance at exact mobile station locations. At the same time statistical measurements in the NMS are gathering information about the uplink field strength levels, quality and handovers. Moreover, these NMS measurements determine the actual traffic (by a base station basis or area basis), congestion, dropped calls and handover failures. After collecting this uplink and downlink information the actual analysis work is commenced in order to understand which planning areas (coverage, capacity or interference) require optimisation.

8.2.1 Coverage assessment

Base station coverage areas are checked and the results are compared with the radio planning predictions. This comparison should indicate if there is a huge difference between the planned and operative radio network. If there are any concerns relating to improvement or requiring verification of the operative coverage the

- coverage limited configuration management, and
- radio plan verification

optimisation tasks are started (network verification).

8.2.2 Capacity assessment

Capacity assessment or analysis often starts from the key performance indicators because the most important statistical results can be found from the NMS which shows the total traffic, congestion (or blocking) and drop calls hour by hour on a base station or area basis. The area values can be used to analyse the usage or efficiency of the deployed base station subsystem (BSS) and the base station values indicate the detailed functionality of the radio network. These base station traffic, blocking and drop call values can and have to be linked to the base station coverage and dominance values before the *capacity limited configuration management* work can be started to determine radio network performance improvements in detail.

8.2.3 Interference assessment

Interference assessment or analysis is started as bad key performance indicators and BER or FER measurements are obtained in the uplink or downlink direction. The purpose of these interference (also called quality) measurements is to find the possible interference areas in the radio network. If the NMS or mobile/PC measurements indicate interference the verification of the whole *radio plan* has to be started because the frequency allocation is very sensitive to radio propagation predictions. When the radio plan has been verified it is time to adjust the frequency allocation and its result. Because frequencies are typically reused as often as possible there is often used a so called "top ten strategy" in a frequency planning assessment. This "top ten strategy" means that, for example every week the ten worst interfered base stations are solved and some improvements are performed in the network in order to improve quality in these base stations. These improvements may cause ten new problem base stations and thus quality problems are not solved but they are moved to a new location. This "top ten strategy" has to be used very carefully.

8.2.4 Functionality assessment

Having established good coverage and enough capacity without interference, calls may still be dropped because of incorrect radio parameter settings which have an effect on the *radio network functionality*. The radio parameters and neighbouring settings are the last detailed part of the optimisation process. The radio resource and mobility management as well as handovers and power control functionalities have to be analysed by using the OMC and mobile/PC measurement results.

8.3 Radio network verification and tuning

The configuration management work (see Figure 8.1) is strongly related to the base station coverage and capacity balance: the larger the base station dominance area the higher the traffic through the base station and vice versa. The aim of configuration management is always to maximise the base station coverage and capacity by taking into account the cost-efficiency and Quality of Service (key performance indicators). In order to optimise the cost-efficiency and the QOS the configuration management have to always be done *over a certain area* (a group of base stations) and the coverage and capacity limited situations have to be separated! In the coverage limited situation the capacity is not an issue at all—typically only one transceiver is required at the base station—and only the radio network coverage has to be deployed with minimum costs. Correspondingly, in the capacity limited situation 2–4 or more transceivers are needed at each base station and this causes coverage limitations, e.g. because of the combiners.

8.3.1 Coverage limited configuration management

The coverage limited radio networks have to be especially planned for rural areas where continuous coverage is required to have a service and where capacity is typically very low. In these areas it is critical to study the coverage over a certain area or coverage over a certain distance, from point A to B as shown in Figure 8.2.

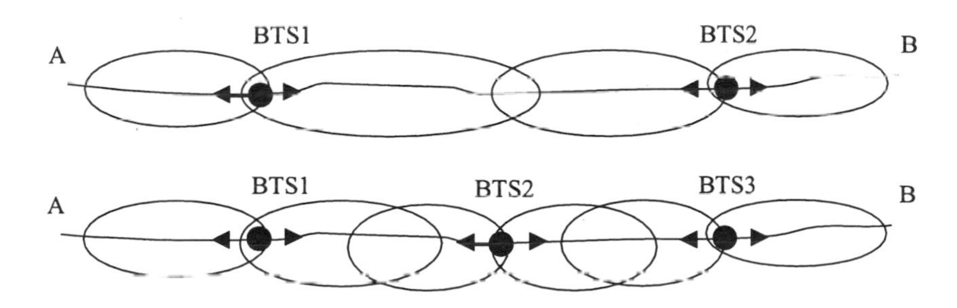

Figure 8.2. Coverage area planning in the coverage limited areas.

Radio planning over a certain area or distance is critical because the aim is to optimise the radio network costs, possible by:

- minimising the number of base station sites (= optimising the base station site configuration)

or

- minimising base station site configurations.

The number of required sites over a certain area or distance depends primarily on the

- regulations in the country (the maximum antenna/mast height, the land purchase/rental conditions)
- available base station site locations (with reasonable work and costs)
- topographic environment.

These primary conditions may cause, for example, that 3 base station site locations (6 cells or sectors) are required to cover the road from point A to point B in Figure 8.2 because of a very hilly terrain environment and available base station site locations. If 2 base station site locations with maximum allowed path loss (a maximum base station configuration) are not able to cover the road, 6 cells are required but not necessarily any more with maximum configuration. It may be enough to have 6 cells without LNA, diversity reception and other base station accessories. Thus, the cost-efficiency of the radio network can be maximised by optimising the base station configurations. The other result could be the reduction in the number of cells down to 4 if new base station site locations could be established and by applying LNAs, diversity reception and boosters in order to utilize the maximum allowed BTS–MS path loss. In this case the cost of each base station site is higher but the overall costs are lower than in the case of 3 base station site locations (remember; the less base station sites the less costs). Not that this work has to be done, of course, in detailed coverage planning but it is also essential during the network evolution when new technologies like GPRS/EDGE/UMTS are introduced (a long-term planning strategy).

8.3.2 Capacity limited configuration management

Capacity limited configuration management has to be done over a certain area as in the coverage limited case. Configuration management in the capacity limited high density traffic areas is not yet as simple as in the coverage limited areas. Coverage and especially indoor coverage has to be considered in these high density traffic areas together with capacity. Thus, the base station dominance area has an essential meaning when base station coverage and capacity are linked. These base station coverage and dominance areas have to be checked first because it has to be established whether the base station dominance areas can be changed without losing indoor coverage. The dominance area sizes have to be modified in order to acquire the traffic at each base station. The dominance areas can be changed by

- tilting the base station antenna
- modifying the base station site configuration
- implementing a new base station or base station site.

If all the neighbour base station dominance areas are maximised and dominance areas can not be changed by configuration management the traffic can be balanced only by implementing a new base station. This is the most expensive solution and thus before applying this solution all other alternatives have to be checked—to rise the antenna height, to move the base station site. When the base station dominance areas have been verified the analysis of capacity and traffic load in the base stations inside the configuration management area can be started; this is demonstrated by an example.

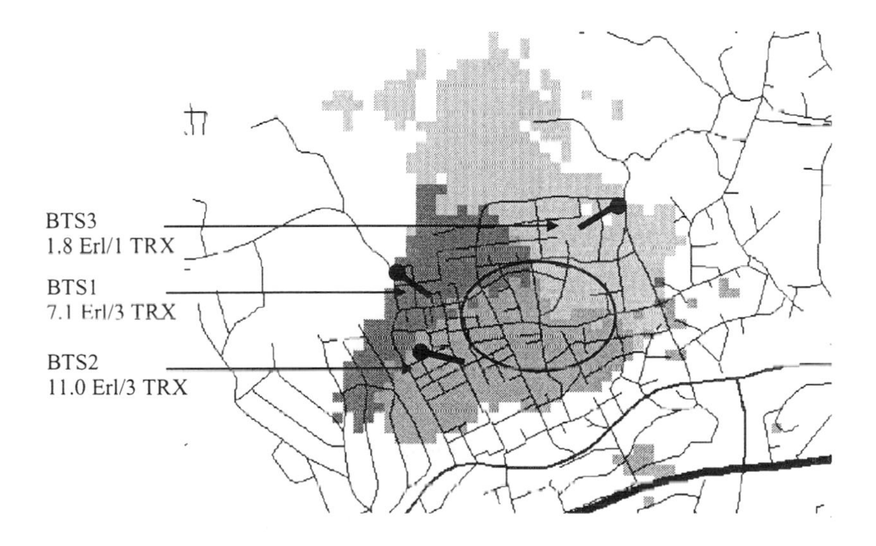

Figure 8.3. BTS1, BTS2 and BTS3 and their traffic load, capacity and dominance areas.

Example: capacity versus dominance areas

Figure 8.3 shows a part of the GSM radio network in a semi-urban medium traffic density area. The configuration management area concerned is between BTS1, BTS2 and BTS3. The base stations' configurations and traffic loads during the two busiest hours over an area (each base station's busy hour is also indicated) have been gathered in Table 8.1.

Table 8.1. BTS configurations and two busiest traffic hours over a day period.

	Erl at 17:00	Erl at 20:00	TRX	Erl at 1% blocking
BTS1	7.1	5.7	3	14.0
BTS2	8.5	11.0	3	14.0
BTS3	1.0	1.8	1	2.5
Total.	16.6	18.5	7	30.5

Figure 8.3 and Table 8.1 show the correlation of the dominance area and the traffic. BTS2 has the largest dominance area and also the highest traffic. Because BTS1, BTS2 and BTS3 are serving the same area and because BTS2 is dominating, the traffic is lower in BTS1 and BTS3. Moreover, only 60 percent (18.5/30.5) of the theoretical capacity of 30.5 Erl is utilised in this area during the *area busy hour* (at 20:00) and thus the radio network cost-efficiency is not maximised. Table 8.1 shows very clearly that the radio network in this area has at least a huge overcapacity which is caused at least by too many TRXs but also maybe by too many base stations because the base station dominance areas are only few hundreds of meters. This conclusion is the end of the configuration management *verification* phase and the next phase is *tuning*.

In the tuning phase it is determined whether

- the number of the base stations can be reduced without losing coverage
- the number of the TRXs in this area can be reduced down to 5
- the TRXs/base stations could be balanced in order to make the frequency planning easier.

For this tuning the field measurements can still be used to define the coverage requirements and the long-term area based traffic forecasts to see if the traffic in the area is increasing slowly or rapidly.

8.3.3 Interference verification and tuning

Interference in the radio interface means problems in the frequency allocation or frequency management and these problems can be caused by inaccurate configuration planning (antenna tilting is not used) or inaccurate propagation models in coverage planning or inaccurate frequency planning tool. Figure 8.4 shows an example of the dominance areas of the neighbour base stations and Figure 8.5 shows the actual coverage area of the BTS 1.

Figure 8.4. Example of the base station dominance areas.

Figure 8.5. Coverage area of the BTS1 in Figure 8.4.

The frequency channel F1 has been reused at both of the depicted base stations and interference has been measured at the indicated locations in Figure 8.6.

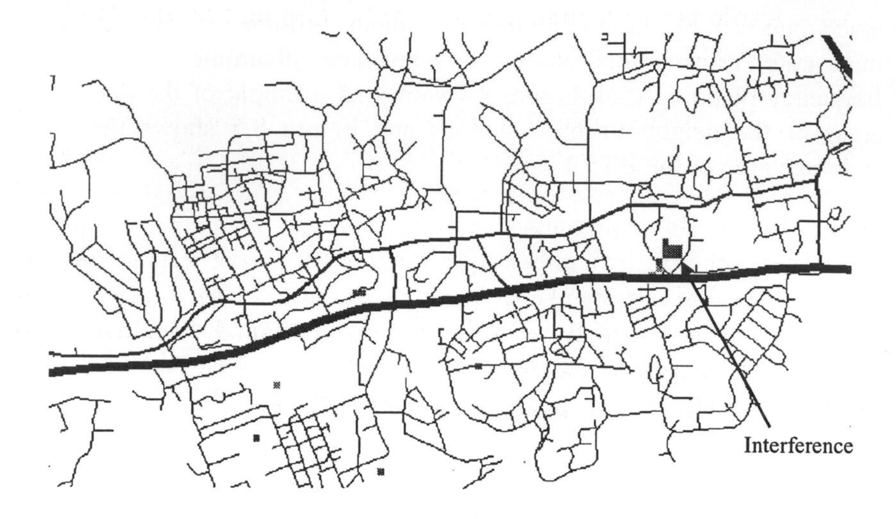

Figure 8.6. Interference between BTS1 and BTS2.

Interference in Figure 8.6 is due to lack of tilting (BTS1 coverage area much larger than dominance area) but it could be because of any inaccuracy in the radio plan like

- quality problems in the digital maps
- errors in radio propagation predictions.

Hence, the digital maps have to be verified and adjusted for the different environments and parameters like

- digital map accuracy and resolution,
- coordination transfer and
- topographic and morphographic digitalisation quality

have to be checked.

Correspondingly, the quality reduction in radio propagation predictions can be caused by

- parameter errors between the planned and implemented base station site configurations
- the errors in the implementation (connectors not joined properly and there is e.g. water in the antenna cables and thus there is an additional loss in the cable because of the implementation error) or
- the change in the propagation environment (new building(s), new residential area, new highway(s)).

8.3.4 Radio network functionality verification and tuning (BSS radio parameters)

The radio network should function correctly and cost-efficiently only if the mobility management can be handled. The mobility management means that a radio connection between the base station and the mobile station has to be maintained even if the mobile station is moving. Moreover, this means that the mobile station has to get the radio service from several base stations because one base station covers only a certain area. These transfers between base stations are called "handovers" and they have to be done based on a certain criteria. These handovers and their criteria are controlled by radio parameters which can be categorised to the classes of:

- signalling
- radio resource management
- mobility management
- measurements
- handover and power control.

The radio network related (base station subsystem, BSS) parameters are defined to control all these radio network functionality related issues in order to know which base station the mobile station should be established on (*signalling*), in order to know where the mobile stations are moving (*mobility management*), in order to know which frequency can be used at different mobile station locations (*radio resource management*), in order to know when the base stations should be changed during the connection (*handovers*), and in order to know how interference could be minimised (*power control*). If these parameter settings are not properly planned and verified the radio network will not work properly even if the radio network is planned properly.

The BSS radio parameters can also be divided into two main categories: *ETSI (European Telecommunications Standardisation Institute) specific parameters* and *base station subsystem supplier specific parameters*. ETSI has specified the general GSM parameters in order to describe the basic functionality of the GSM radio system. In addition to this all equipment suppliers can have their own software features and parameters compatible with the ETSI GSM specifications. The supplier related parameters are typically innovations to improve signalling, mobility, radio resource, and handover or power control. These supplier related features and parameters can not prevent any other basic functionality and they typically have to be independent of the mobile station type in order to have wide applicability. All radio parameters also typically have "default" values that are average values, based on the long-term experience of different areas, and radio network parameterisation is usually commenced with these default values; slightly different parameter values are required for *urban*, *rural* and *indoor* environments.

Different radio parameters are always related to certain radio network functionality that is controlled by measuring a specific value (e.g. field strength level, bit-error-rate, mobile station distance from the base station). Signalling, mobility and radio resource management parameters, which are not strongly related to the environment (they do not have a great influence on the base stations' dominance areas), are typically quite constant and their values depend more on the base station configurations or on the whole radio network configuration that is defined in the detailed planning phase. When the radio network configuration is confirmed almost all of these parameter values can be defined and they also can be frozen or locked. Moreover, handover and power control parameters depend strongly on the environment and they can be used to optimise the base station's dominance area. The aim of radio planning is to use certain base stations on certain coverage areas and also certain frequencies on certain areas. The purpose of the handover and power control parameters is to ensure that the base station's dominance areas are equal with the existing plan. An unhomogenous environment—as with buildings, road or street canyons—may cause extraordinary situations in the field and thus the handovers from the base station to the other may fail. These handover failures may be caused, for example by the interference peak at a certain location, and thus handovers should be made earlier or quicker. By changing the handover or power control parameters the base station dominance area borders can be optimised to avoid handover failures and furthermore drop calls. This is a typical verification and tuning task that is related to the network functionality and radio parameters in the different environments.

8.3.4.1 BTS ⇔ MS signalling

Signalling here means the radio interface signalling between the BTS and the MS in the uplink and downlink directions and it is basically required to inform the mobile station about the radio network configuration and to inform the base station about the mobile station position in the field. In order to manage this discussion both in IDLE (there is no call) and DEDICATED (there is a call) modes signalling channels, shown in Table 8.2, are needed.

Table 8.2. Signalling channels in the uplink and downlink directions.

BTS→MS			MS→BTS		
BCCH		Broadcast Control CHannels	**BCCH**		Broadcast Control CHannels
	FCCH	Frequency Correction CHannel			
	SCH	Synchronization CHannel			
	BCCH	Broadcast Control CHannels			
CCCH		Common Control CHannels	**CCCH**		Common Control CHannels
	PCH	Paging CHannel		RACH	Random Access CHannel
	AGCH	Access Grant CHannel			
DCCH		Dedicated Control CHannels	**DCCH**		Dedicated Control CHannels
	SDCCH	Slow Dedicated Control CHannel		SDCCH	Slow Dedicated Control CHannel
	SACCH	Slow Associated Control CHannel		SACCH	Slow Associated Control CHannel
	FACCH	Fast Associated Control CHannel		FACCH	Fast Associated Control CHannel
TCH		Traffic CHannels	**TCH**		Traffic CHannels
	TCH	Traffic CHannel		TCH	Traffic CHannel

Note that a "channel" does not have a physical meaning here and these connections, where information is transferred between the base station and mobile station, are called channels but are definitely not frequency channels or time slots (but are a part of time slots). The physical meaning of the signalling "channel" is explained later.

BTS→MS (IDLE/DEDICATED)

The signalling is required in the downlink direction for mainly informing the mobile station about the radio network configuration.[1] The *broadcast common channels* (BCCH) and *common control channels*

(CCCH) are used for this purpose. The BCCH contains information about the radio network and is thus critical. If the BCCH information can not be decoded by the mobile station it may lose the information about essential parameters from the network. The CCCH are used for paging (PCH) and informing (AGCH) the mobile station which slow dedicated control channel (SDCCH) the mobile station can use before the traffic channel (if SDCCH is required in call establishment). Thus, the BCCH is sent in both idle and dedicated mode and the PCH triggers the dedicated mode when the question is about mobile terminated calls (MTC). The AGCH is used in the dedicated mode just after the PCH and the random access channel (RACH) which corresponds to the PCH in the uplink direction.

The *dedicated control channels* (DCCH) like the slow associated control channel (SACCH), fast associated control channel (FACCH) and slow dedicated control channel (SDCCH) are not used continuously but only when needed. Power control commands are sent on the SACCH and there is some space left to send short messages if required. The FACCH is used for the handover commands and this FACCH is actually a normal traffic channel (TCH) with a flag which says that a TCH is used for signalling at that moment. The FACCH is thus as fast as the TCH what means that it is many times faster than the SACCH. Therefore, there are no actual delays or delay problems during handovers. The SDCCH is used primarily for short messages, location updates and the call establishment process. The channel capacities and their requirements are dealt with later but it can be noted here that the critical signalling channels in the downlink direction are the PCH, AGCH and SDCCH due to the capacity and the BCCH due to the interference.

MS→BTS (IDLE/DEDICATED)

The uplink direction has almost the same channels as can be found in the downlink direction but the usage of them is slightly different. There is only one common control channel in the uplink direction called the random access channel (RACH) which always starts the communication between the MS and BTS. If the call is mobile originated (mobile user makes a call) the RACH is sent first in the uplink to indicate a need for a radio channel. The system responses by sending an AGCH which contains information which SDCCH (frequency, time slot) the mobile station can use in the next uplink transmission. On the SDCCH the mobile station sends a service request (e.g. call establishment (speech/data call), location update), the authentication and ciphering are done and finally the connection is ready for the speech or data. If the mobile receives a call (a mobile terminated call) the procedure is the same except the downlink paging on the PCH starts the process and then the RACH follows.

In the uplink direction the SACCH is fully occupied (there is no space for anything else) sending the mobile station's measurement results (from the neighbour base stations) to the serving base station. It has to be noted that these measurement results are reported only in the dedicated mode and in the idle mode there is no connection between the mobile station and base station (mobile station only receives information from the network). The power control as well as handover commands are sent on the FACCH in the uplink direction because the SACCH is full. Moreover, the SDCCH is used as in the downlink direction.

Channel structures

How and when these signalling channels are used in the downlink and uplink directions is found in the channel structure which also indicates the capacity limitations for each channel. The description of the channel structure starts from the time division multiple access (TDMA) principle to explain how the frequencies are used (recall Figure 1.4). Initially, all frequency channels are divided into 8 blocks (0–7) called time slots which use the same frequency periodically and which together define a TDMA frame.

These time slots are used to send either traffic (voice/data) or signalling (messages/commands) and the signalling time slots are usually used on the first frequency (called BCCH frequency). The number of signalling time slots depends on the signalling capacity need and the remaining time slots can be used for the speech/data applications. The signalling channel requirements as well as the signalling channel structure can be divided into two groups: the BCCH/CCCH and SDCCH. Both these groups (meaning all the signalling channels) can be used on the first time slot (time slot 0) on the BCCH frequency which means that there is a limited capacity on the PCH/AGCH/SDCCH but 7 time slots can be used for the speech/data traffic. If this configuration does not provide enough signalling capacity, the BCCH/CCCH and SDCCH can be separated to different time slots and thus these configurations are called combined and separated (recall Figure 6.2), respectively.

As mentioned the combined configuration provides one more time slot for the traffic but decreases the signalling capacity and the separated configuration does the opposite. The required channel configuration depends typically on the SDCCH and PCH capacity needs which are studied next by focusing initially on the detailed signalling channel structure on the time slot zero in the combined and separated configurations. The signalling channels in the combined signalling configuration on the time slot zero are shown in Figure 8.7 in the downlink and uplink direction.

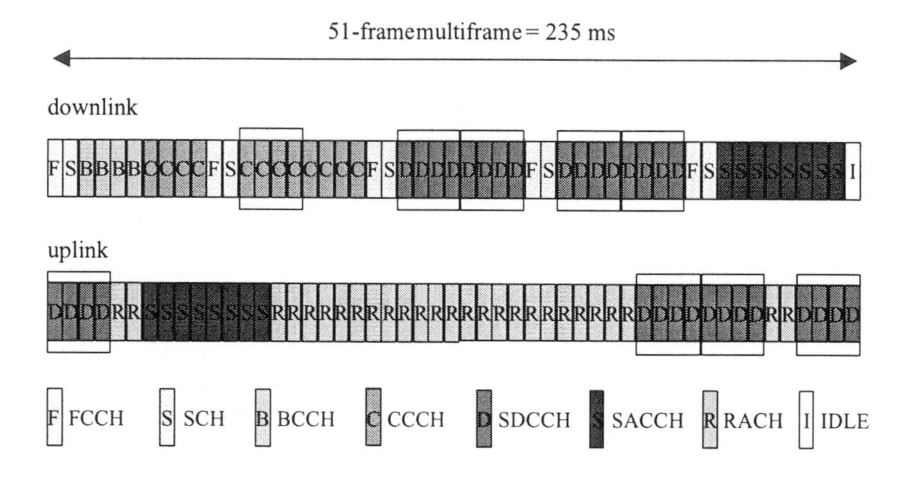

Figure 8.7. The BCCH/CCCH/SDCCH blocks in the combined configuration on the time slot zero (ts0).

Note that there are only 3 CCCH blocks (a block equals 4 consecutive time slots which is a full message) and 4 SDCCH blocks in the combined configuration in the downlink direction which is critical because the CCCH blocks still have to be divided between the PCH and AGCH. Figure 8.7 also shows that one time slot (ts0) is used for the several signalling "channels" and thus the signalling channels are called *signalling logical channels*. Each signalling logical channel (a block) performs the signalling procedures of one mobile station and thus time slot zero is used for the signalling of several mobile stations. If the combined configuration is compared to the separated configuration that is presented in Figure 8.8 and Figure 8.9, it can be seen that the only difference is in the number of the CCCH and SDCCH blocks (9 CCCH blocks on the ts0 and 8 SDCCH block on the ts1).

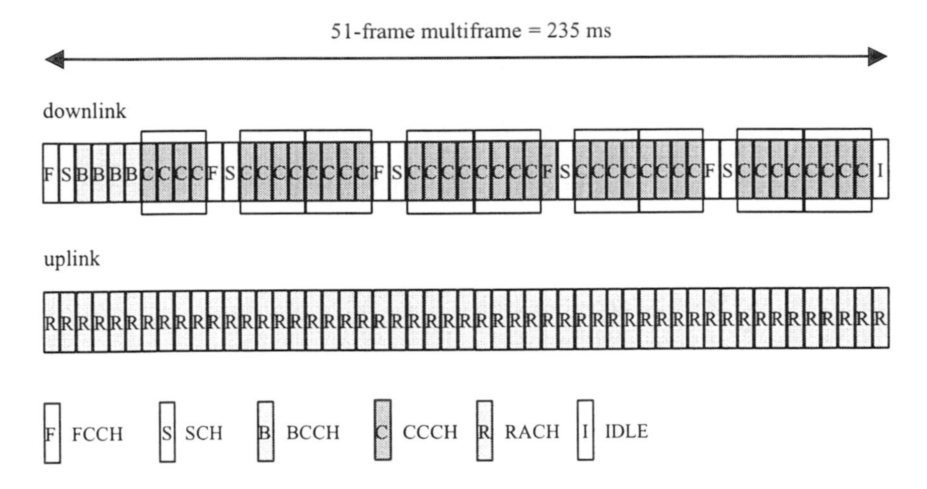

Figure 8.8. The BCCH/CCCH configurations on the ts0 in the separated configuration.

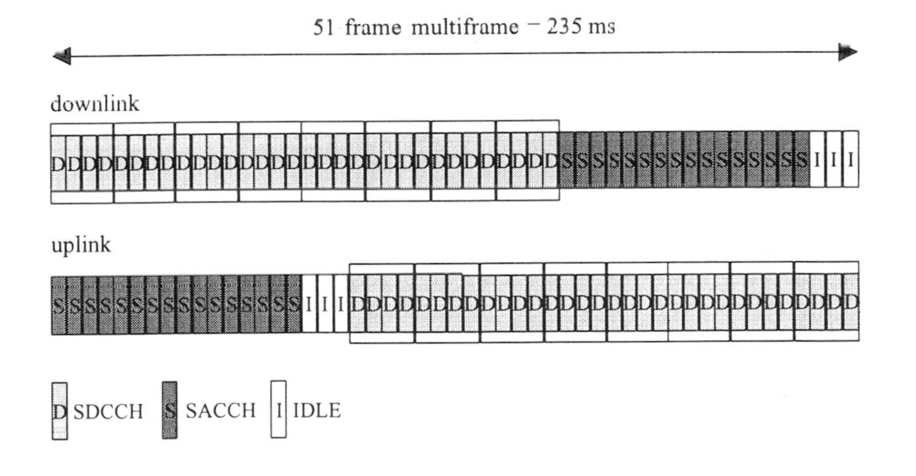

Figure 8.9. The SDCCH configurations on the ts1 in the separated configuration.

Because it has been mentioned several times that the PCH/AGCH and SDCCH capacities are in the key role when defining the signalling parameters it is now time to take examples about these signalling channels.

Example of Paging CHannel (PCH)

This example concerns

- the combined BCCH/SDCCH configuration
- because it may limit the paging capacity due to only three CCCH blocks for the PCH and AGCH purposes. In this combined configuration
- one CCCH block is reserved for the AGCH and 2 blocks for the PCH and
- 4 mobile stations can be paged per paging channel block when the temporarily mobile subscriber identification (TMSI) is used
- and it can be calculated straightforwardly that
- 4 pages/blocks * 2 blocks = 8 pages every 51-frames multiframe (235 ms, see Figure 8.7),
- and finally ((3600 * 1000) / 235) * 8 > 120000 paged MS per busy hour.

This means that 120000 mobile stations (or mobile terminated calls) can be paged during a busy hour at every base station where 2 CCCH blocks are reserved for the PCH. This figure indicates only that there can be made approximately 120000 *mobile terminated calls* (typically around 70 percent of all calls are mobile terminated) during a busy hour. This paging capacity is required for the whole location area and the maximum location area size (number of base stations) can finally be estimated when the total traffic and the ratio of mobile originated and terminated calls in the area are known. Also some margin have to be left for re-paging.

Example of SDCCH

This SDCCH example considers mostly the influence of the call establishment and the location updates (note location area borders) on the SDCCH capacity. The short message load can be studied from the network and is not included as it can vary a lot. This example shows that the location updates load the SDCCH excessively and thus the location area borders have to be planned carefully. The call establishments load the SDCCH based on the

- base station configuration: 2 TRXs / cell ~ 8.11 Erl / cell (1 percent blocking probability)
- average traffic per subscriber: 1.5 min / subs / BH = 25 mErl / subs

- number of subscribers: 8.11 Erl / cell /25 mErl / subs = 325 subs / cell
- authentication and ciphering = 4 s = 1.1 mErl / call (SDCCH reservation time)
- → 325 calls / cell * 1.1 mErl / call = 0.3575 Erl / cell (SDCCH)

and correspondingly the location updates (time periodic only) requires the following SDCCH reservation:

- each mobile station makes a location update once in 60 minutes (this is based on the parameter settings)
- 325 subs / cell
- SDCCH reservation time for location update = 4 s = 1.1 mErl
- → 325 calls / cell * 1.1 mErl / call = 0.3575 Erl / cell (SDCCH)

When these SDCCH reservations (in Erlangs) are calculated together and studied with the SDCCH channel requirements in Erlang-B table (see Table 6.1) it can be noted that 4 SDCCHs are needed for this configuration (each subscriber makes one call per busy hour and one location update) and thus the combined channel structure is enough and there is still some space for the short messages. This SDCCH example, as well as the previous PCH example, shows the capacity limitations of these signalling channels and indicates the purposes of the signalling parameters.

8.3.4.2 Radio resource management

Radio resource management concerns the management of the radio connection between the base station and mobile station. This management has to be done when the mobile station is in the IDLE or DEDICATED mode and can be divided into two main parts; IDLE mode control and DEDICATED mode control. In both these modes the mobile station has to be informed of the *access* to use the base stations and about the usage of the *radio resources* of the base stations. Simultaneously, the radio system controls the access and the radio resources that are used by the mobile station. The access management in both IDLE and DEDICATED modes contains a group of IDs and a group of parameters to define when the base stations can be used (the actual "access parameters"). Correspondingly, the radio resources are controlled by a group of parameters to define the best radio quality for the BTS–MS connection. Thus, the radio resource management parameters really control the IDLE mode (first the IDs and then the actual access) and DEDICATED mode (the access, the radio resource allocation and release) and the most important parameters related to this mobile call process (IDLE mode, BTS access, DEDICATED mode, radio resource allocation, call release) are explained in this order.

IDs (IDLE mode)

Different IDs are required to separate different radio networks from each others. Frequencies separate different operators in the same country but in country border areas the same frequency bands may being used and thus different IDs are definitely required. Radio network IDs also help the mobile station to recognise the base stations of the own network which is critical for billing and in special situations such as emergency calls. Moreover, radio network IDs can be utilised in the gathering of statistical data because the results can be connected to the location areas or to certain base stations. The essential and ETSI specific IDs are presented in Table 8.3 that shows two main parameterisation items called base station identity code (BSIC) and location area ID (LAI).

Table 8.3. The most essential IDs to define and separate the base stations of the radio network.[2]

BS identity code	
Network colour code	0–7
BTS colour code	0–7
Location area id	
Mobile network code	244
Mobile network code	5
Location area code	xxx

BSIC is actually the combination of the base station colour code (BCC) and network colour code (NCC) and can be calculated by using modulo 8 method meaning that the value of the BSIC = 8*NCC + BCC. This BSIC value is needed to separate the co-channel frequencies from each other when mobile stations are measuring the received power levels of the neighbour base stations in the downlink direction (see Figure 8.10). These measurements are performed on the BCCH frequencies and the results are used to select the best base station (the best server) in the IDLE mode and to evaluate and trigger handovers in the DEDICATED mode.

The location area ID (LAI) is another critical group of IDs containing mobile country code (MCC), mobile network code (MNC) and location area code (LAC). The MCC and MNC are defined by the MoU (Memorandum of Understanding between operators) and LACs have to be defined during the planning process. The same LAC has to be used for the same location area and thus it is critical to use correct LAC values in order to avoid unnecessary location areas and location updates. The base station IDs are not mentioned in Table 8.3 but they are of course also needed for each base station.

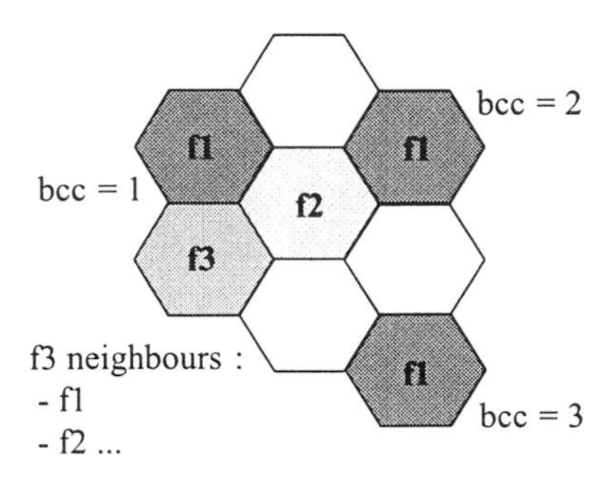

Figure 8.10. The need of BSIC for the frequency co-channel management.

Access (IDLE mode)

When each base station has its own IDs (BTS ID, BSIC, LAI), the access (or access control) to these base stations at the different mobile station locations has to be defined. Access definition begins with the parameters which define whether use of the base station is permitted. It is critical to know [2]

- the frequency
- whether the base station is barred or not
- whether call re-establishment is allowed or not
- whether the emergency calls are permitted or not
- whether the specific mobile station classes are not allowed use at the base station.

Table 8.4. The essential base station access parameters.

Frequency band in use	gsm
Cell barred	no
Call re-establishment allowed	no
Emergency call restricted	no
Not allowed access classes	
PLMN permitted	0–7

Typically, these parameters have the fixed values which are presented in Table 8.4. The last parameter in Table 8.4, PLMN permitted, controls the mobile station measurements. The mobile station measures the frequencies that are required by the system but the mobile station reports and utilises only the measurement results which have the correct PLMN permitted value. Thus the PLMN permitted parameter is critical e.g. in the country border areas. All these above mentioned parameters define the permission to use the base station and precede the IDLE mode access parameters, that control the radio link (define the best server).

Table 8.5. The essential base station radio link control parameters.[3–4]

rx_level_access_min	-105
cell reselect hysterisis	6
ms_tx_power_max_cch	33

In the IDLE mode (as in the DEDICATED mode) it is to be defined whether the base station can be used and which base station is the best server. First, the parameter rx_level_access_min (see Table 8.5) defines the minimum level that the base station can be used in the call establishment. Next, this same parameter and ms_tx_power_max_CCH (mobile station's maximum transmission power in call establishment) are used in the equation C1 to define the best server

$$C1 = (A - \max(B,0))$$

A = averaged receiving level – p1
B = p2 – maximum RF output power of the mobile station
p1 = rx_level_access_min
p2 = ms_tx_power_max_CCH.

The received power levels from the defined neighbour base stations are measured by the mobile, C1 values are calculated (by the mobile), C1 values are compared (by the mobile) and the base station with the highest C1 is selected as the best server every 5 s. If the base station has 1 dB better C1 value than the others it is the best server excluding the location area border where hysteresis are used in order to avoid too many location updates. This hysteresis is defined by the cell_reselect_hysteresis [3] parameter and it is used when comparing the C1 values between the base stations from different location areas. The base station from the new location area is selected as the best server only when it is e.g. 6 dB better.

The more advanced base station selection called C2 can also be applied for example for multilayer (dual band) networks when one layer is wanted to be prioritised to have call establishment. The network has to indicate to the mobile station by parameter cell_reselect_param_ind [3] whether it has to use C2 procedure. Simultaneously, the network sends the parameter values to the mobile station to calculate the C2 values, using the equations:[3]

$$C2 = \begin{cases} C1 + \text{cell\_reselect\_offset} - \text{temporary\_offset} * H(\text{penalty\_time} - T), \\ \quad \text{penalty\_time} <> 11111 \\ C1 - \text{cell\_reselect\_offset}, \text{ penalty time} = 11111 \end{cases}$$

$$H(x) = \begin{cases} 1 \text{ for } x \geqslant 0 \\ 0 \text{ for } x < 0 \end{cases}$$

A basic example of using C2 is to set the parameter penalty_time < > 11111 and temporary offset = 0 dB what means that equation C2 = C1 + cell_reselect_offset. If the value of the parameter cell_reselect_offset is for example 10 dB the base station has 10 dB higher C2 value and it is 10 dB more favourable.

Access/IDs (DEDICATED mode)

Base station selection is the last phase in the IDLE mode control before the mobile originated or mobile terminated call (MOC/MTC) establishment procedure. When these procedures have started it can be said that the mobile station is in DEDICATED mode and more control is required. First of all, the mobile station still needs all the same IDs as in the IDLE mode and in addition BCCH frequencies of the neighbour base stations in order to measure the right neighbour base stations. All this information is sent to the mobile station on the system information messages on the SACCH signalling channel.

The main concern of the radio resource management in the DEDICATED mode is to manage the radio resource control as frequencies and time slots and the IDs are not concerned too much because they are the same as in the IDLE mode. The key areas in this radio resource control are the traffic channel allocation and the influence of the extra features such as frequency hopping. The basic idea of traffic channel allocation is based on the minimisation of interference in the downlink and uplink directions. This can be controlled only in the uplink direction by the base station by measuring the free time slots (time slots which have no traffic). When the base station measures the time slot without traffic the measurement result has to be noise or interference (from the mobiles using the same channel at another base station) because there is no traffic. The measured power levels are divided into different categories by parameter and the time slot with the lowest interference level can be selected. The frequencies are also selected by using different algorithms as for example to prefer BCCH frequencies first or to use all frequencies in turn.

When the frequency and the time slot are selected the base station (and mobile station) has to know how to use them and this depends on the applied frequency usage related features as frequency hopping that contains some more detailed parameters.

8.3.4.3 Mobility management

Signalling takes care of the conversation between the network and the mobile station and this conversation mainly contains the key information about the network. Radio resource management mainly contains the usage of the radio elements, as frequencies and time slots. Correspondingly measurements and handovers which are strongly related to radio resource management are considered as separate topics. Mobility management exists in all these areas and has to be generally understood and highlighted as key aspect regarding the mobile location in the network in IDLE and DEDICATED mode.

The mobility management can be considered as the radio resource management by starting from the IDLE mode, analysing the DEDICATED mode, and finally concerning the functions related to call release. In the IDLE mode the network only knows the location area where the mobile station is and uses this information for paging. This is the reason why the mobile station has to inform the network every time it changes location area. Thus the C1 or C2 parameters and the cell_reselect_hysteresis [3] could also be related to the mobility management. In addition the time periodic location update is used to ensure that the mobile station is still in the network and to ensure that the information about the mobile station in the network is still correct.[1] When the mobile station makes a call or receives a call it starts the DEDICATED mode and the network knows the mobile station location in the base station level because the network (master) controls the radio resources and the mobile station functions based on the orders coming from the network (the mobile is slave); for example the network orders the mobile station to perform a handover to the certain target base station. Thus, the measurements and handovers and their parameters can be related to the mobility management (or to the radio resource management). When the mobile drops or finishes the call, the call drop (abnormal release) or normal call release occurs. After releasing the connection the mobile station has to update the location if it has changed during the call. If the location area is the same as at the beginning of the call the mobile station starts to analyse the C1 or C2 criteria. In this IDLE mode there is still the IMSI_attach_detach [2] parameter which controls whether the mobile station is switched on or off. The mobile station sends IMSI attach or detach automatically (if the function is used) to the network to inform that it is in the network or not. These are the main parameters and functions to control the mobile stations movements in the IDLE or DEDICATED modes.

8.3.4.4 Measurements and measurement processing

Measurements are required to control the radio resource and mobility management, meaning that the network has to offer the best server (the best base station equals the best frequency and the best overall quality) for the mobile station at different locations. If the neighbour base station has better quality or better power level the handover to that base station has to be done based on the handover criteria. In addition to these actions the power levels of the mobile stations and base stations can be controlled in order to decrease interference. This power controlling is also based on the measurements done by the base stations and mobile stations.

Measurements are done in the same way in the uplink and downlink directions and thus it is enough to consider the mobile station measurements because these contain some limitations which do not happen in the uplink direction (the base station has enough time and capacity to do the measurements almost as necessary). The mobile station reports (sends) the measurement results of the six best neighbour base stations and the serving base station every 480 ms. During this SACCH period the mobile station measures the neighbour base stations (maximum 32 neighbours) and decodes the BSICs of these base stations. The measurement results are not reported to the base station if the BSIC is not decoded or if the measurement results are coming from the base station which has an incorrect PLMN permitted value.

After measuring the uplink and downlink directions the measurement results are gathered to the base station where the pre-averaging is done if required. After pre-averaging the base station starts to average the measurement results or the results are sent to the base station controller where they are averaged.

The averaging process (for the uplink and downlink directions) is done for the bit-error-rate (called quality in the GSM) and the received power level measurements. The parameters p, n and window_size are critical [3] to achieve the appropriate results. The window_size defines how many results are averaged and the p and n values how many averaged results (p) have to exceed e.g. the handover threshold inside a certain window (n).

P (dBm)	FS (dBμV/m)	LEV	BER (%)	BER (%)	QUAL
–110	27	0	**Range**	**Mean**	
–109	28	1	< 0.2	0.14	0
–108	29	2	0.2–0.4	0.28	1
.	.	.	0.4–0.8	0.57	2
.	.	.	0.8–1.6	1.13	3
.	.	.	1.6–3.2	2.26	4
–49	88	61	3.2–6.4	4.53	5
–48	89	62	6.4–12.8	9.05	6
–47	90	63	> 12.8	18.1	7

Figure 8.11. The units of power and quality levels in the BTS–MS measurements and measurement processing. [3]

Figure 8.11 shows the values of the received power and quality measurements. Note that decibel units are used in the power level measurements and correspondingly the mean BER values (the values 0.14–18.1) are used when calculating the quality results. The last value called QUAL is a GSM quality class that is used in handovers and power controls.

These units have a strong effect on the averaging strategy and the following example demonstrates the inaccuracy of quality level measurements. The measured values (GSM quality classes) are

0 (the mean used in the calculation is 0.14), 0, 7 (the mean is 18.1) and 0.

The averaged value is straightforward

$$(0.14 + 0.14 + 18.1 + 0.14) / 4 = 4.63$$

which corresponds to the quality class 5 (see Figure 8.11) which furthermore would cause a handover if a typical handover threshold 4 were used in the network. Hence, only one bad sample (quality class 7) would maybe cause an unnecessary handover that may deteriorate the overall quality (functionality) in the network. Thus, the averaging of the quality samples does not necessarily correspond to the situation in the field.

8.3.4.5 Handovers

Handovers have to be performed in order to offer the best radio quality to the mobile stations at the different locations in the network. The decision to make a handover is made based on the uplink and downlink power level and quality measurements and final handover triggering happens by using the handover algorithms in the base station and/or base station controller which command the mobile station to the new target base station. The handovers can be divided into a few main groups called

- power level handovers (uplink and downlink)
- quality handovers (uplink and downlink)
- better base station handovers—power budget and other supplier based handovers.

All these handovers contain the same process with a few phases which have to be understood in order to understand the usage of the parameters for these different phases. The handover process includes

- the measurement result and the handover threshold comparison
- the evaluation of the six best neighbour base stations
- the target base station selection

phases. The handover process begins after the averaging process with the *measurement result* and *handover threshold comparison* as presented in Figure 8.12.

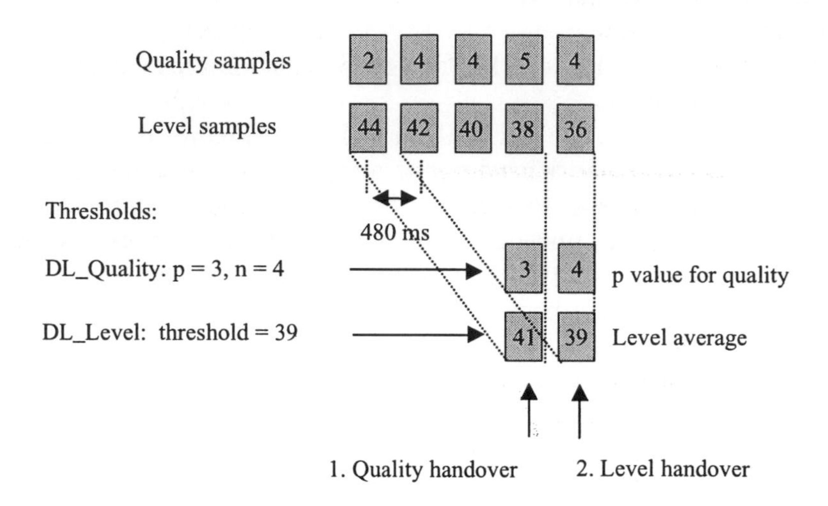

Figure 8.12. The comparison of the measured and averaged results.

The measurement result and the handover threshold comparison

This comparison is made between the serving and the neighbouring base stations by utilising the supplier based handover algorithms. The first purpose of this comparison is to solve when the different handovers are triggered (handover algorithms are required for all types of handover) and second to solve which neighbour base stations fulfil the conditions of function as a new serving base station (the evaluation of the six best neighbour base stations).

The first part of the handover threshold comparison—the handover triggering—begins by studying which handovers are triggered in the uplink or downlink directions after each new measurement result (after each SACCH period). All different handover types can be checked after each SACCH period and the averaged results are used for this purpose. In Figure 8.12 the last power level sample of the serving base station in the downlink direction is 36 (in the GSM units), the averaging window is four samples and thus the averaged value after the last sample is 39. The handover threshold for the downlink in this example is 39 for the power level and thus at this point the power level handover would be triggered if $p = 1$ and $n = 1$. Correspondingly, the last four downlink quality samples of the serving base station are four, four, five and four (note that averaging is not used for the quality samples). The quality handover is triggered after sample value 5 if the typical threshold value four would be considered with the values p = 3 and n = 4 and if averaging is not used. The same comparison for the power levels and quality has to be made also in the uplink direction. Additionally, it has to be remembered that the better base station handovers like power budget handovers also have to be checked. If the different handover types are triggered at the same time the priority of the handover types would define which handover type would be used as a reason to make a handover. The handover priorities are typically classified as

- uplink/downlink quality
- uplink/downlink level
- better base station (power budget and others).

When the thresholds have been compared and the handover type (reason) has been solved it is time to define which neighbour base stations fulfil the conditions of being a candidate target base station. This candidate base station evaluation is done by using the supplier based handover algorithms.

Neighbour base station evaluation

The neighbour base station evaluation is a process where all the best neighbour base stations are compared to the serving base station in order to decide whether each neighbour base station is able to provide better service (overall quality: coverage, capacity and BER) than the actual serving base station. This evaluation is done based on the supplier based handover algorithms and the result is the group of candidate base stations for the handover target and the final target base station is selected from this group of candidates.

8.3.4.6 Power control

Power control is the other function that utilises the uplink and downlink measurement results to control the base station and mobile station power in order to decrease interference and save power consumption of the equipment in the radio network. The aim is to make the power control before the handovers because handovers should always have higher priority than power controls. It also has to be remembered that on BCCH frequency the power control can not be used in the downlink direction because continuous transmission is required.

The principle of power control is to define an upper limit and lower limit and the received power and quality level should be maintained inside of this window in order to optimise power consumption and quality in the radio network. Figure 8.13 shows this principle of power control and power control parameters in case of the received level. It can be seen in Figure 8.13 that parameters are needed to define the upper and lower limits and, of course, the time period when triggering happens.

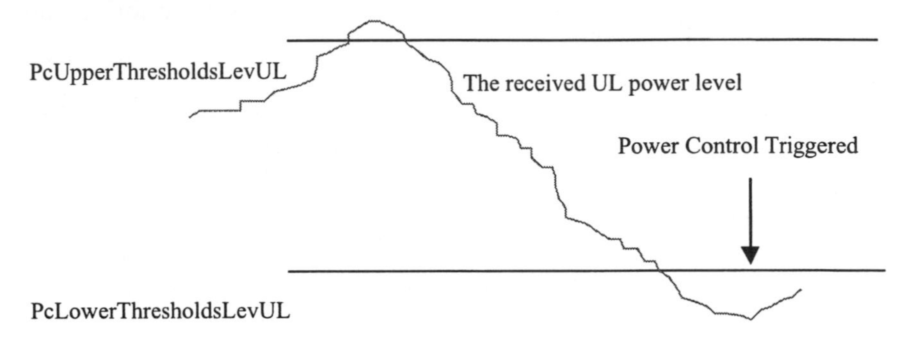

Figure 8.13. The power control process.

This window is typically set by 10 dB hysteresis (the difference between the upper and lower limits) which may be even too high. The window of 4 dB optimises the received power level more accurately and may provide even better radio quality but it also loads the system much more. However, there are no system limitations (e.g. signalling) that primarily prevent such a fast power control.

8.4 Conclusions

The topics of optimisation (of the radio network configuration and a short explanation of the most important radio parameter areas) led to an understanding of the optimisation process and the observation of two key issues of radio network optimisation: configuration management and radio network functionalities (see Table 8.6).

Table 8.6. Optimisation.

Subject	**Findings**
Radio network assessment	–Measure the uplink and downlink power levels and BER and FER –Coverage area check
Configuration management	–Traffic over the area –Traffic/base station –Dominance areas of the base stations
Radio plan quality	–Digital maps –Radio prediction model –Accuracy of the radio prediction –Frequency allocation
Radio network functionality	–Signalling parameters –Radio resource parameters –Measurements and measurement processing –Handover parameters –Power control parameters

8.6 References

[1] M. Mouly, M-B. Pautet, "The GSM System for Mobile Communications," 1992.

[2] ETSI, Digital cellular telecommunications system (Phase 2+), Function related to Mobile Station (MS) in idle mode and group receive mode, GSM 03.22.

[3] ETSI, Digital cellular telecommunications system (Phase 2+), Radio subsystem link control, GSM 05.08.

[4] ETSI, Digital cellular telecommunications system (Phase 2+), Radio transmission and reception, GSM 05.05.

Chapter 9

RADIO NETWORK MONITORING

9. RADIO NETWORK MONITORING

The last and one of the most essential parts of the radio network system planning process is monitoring. Monitoring is a key function in learning about actual radio network performance and in finding correct radio network assessment areas to be focused in order to improve the radio interface Quality of Service (QOS). Monitoring has to be done continuously and it has to be based on radio system planning targets in order to support the achievement of these targets. Long-term monitoring results can be used as an input value for dimensioning work (for the radio network extensions) and short-term results for optimisation work. Moreover, monitoring can be divided into *radio network functionality* and *radio network configuration* monitoring. Network functionality monitoring is performed by defining the key performance indicators (KPI) which tell how efficiently the radio network is used and what the QOS of the radio interface is. Configuration monitoring concentrates on managing the radio network infrastructure and finding its essential bottlenecks. Both of these monitoring types also need software applications in order to be able to show the possible assessment areas.

9.1 Radio network functionality

Radio network functionality is the first monitoring part and highlights a good connection between the mobile station and base station. This connection means actual good quality of speech or high data transmission rate both of which can be monitored by using the key performance indicators.

9.1.1 Key performance indicators

Key performance indicators (KPI) are needed to measure *cost-efficiency* and *QOS* of the radio network and to show the planning areas which possibly need assessments. These KPIs are divided into these two main groups which both include different KPIs to inform the performance of the radio network. The radio network cost-efficiency can be observed by measuring the *effective usage of the network* and *frequency bandwidth* and by defining certain KPI values to evaluate these topics. Moreover, a good radio network QOS requires high *call success rate* which means that the user has a *successful call attempt*, *good voice quality* and *normal call release*. In order to analyse the radio QOS and to understand the radio network performance in detail, the KPI values such as *call success rate*, *handover success rate*, *drop call rate* and *blocking* have to be measured. Each of these cost-efficiency and QOS key performance indicators can be measured in different ways, and thus, a consistent measurement method has to be defined.

9.1.1.1 Radio network cost-efficiency

Cost-efficiency can be analysed by measuring the usage of the radio network and frequency bandwidth. Usage of the radio network can be solved by measuring the total traffic over a certain area (containing a certain number of the base stations) and comparing this value to the theoretical capacity that is implemented in the radio network (= the total traffic that can be offered by the implemented transceivers based on the Erlang tables). It is important to know the network configuration and the number of the traffic time slots exactly in order to receive correct information about the network status. In addition, it is required *to measure the traffic and usage over a certain area*, containing several base stations in order to get an understanding of changes in the total traffic and usage. If only an individual base station is observed, which does not gather traffic today and which is congested after six months, it is not possible to conclude whether this base station is working efficiently.

The aim of radio system planning is the planning of a radio network with a required capacity for a certain area and simultaneously offering a required coverage. If coverage is not a problem, approximately *80 percent of the radio network capacity* should be used before capacity extensions are installed. This is an example about an urban area while differing monitoring targets have to be defined for a rural area where base stations are mainly implemented due to lack of coverage. Optimised cost-efficiency can be achieved in rural areas by increasing antenna heights so that base stations can cover such an area that contains enough traffic and that could be served over the next five years while taking into account the traffic increase forecast. In practice this kind of strategy is typically not possible because of limitations by the government for maximum antenna height and lower target value while *35–50 percent for the usage are typically accepted in rural areas*. Antenna heights between 100–120 m are also avoided by claiming that high antenna positions cause interference to urban areas which is actually false. Interference can be avoided by tilting and increasing antenna height in steps (35 ... 50 ... 75 ... 100 m) when moving away from the city.

When measuring the usage of the radio network, the maximum traffic need has to be found for the measured area. A value that could be called an *area busy hour*. The area busy hour usually has to be addressed by tailor-made functions as commercial database programs do not *necessarily* support this kind of analysis. NMS measurements and post-processing tools are still required in order to gather raw data that can be analysed. The area based measurement takes into account the *short-term* daily changes, when people move for instance between their homes and offices, and *long-term* changes, for example, new residential and office buildings. Short-term changes cause the maximum usage of the 80 percent in urban areas and thus 20 percent is reserved as a margin for the daily changes in the urban area. This typical value of 20 percent daily change can be obtained by calculating the total traffic from a certain area by using

- busy hour values from each base station (the highest traffic at each base station during a day period) and
- area busy hour value (the highest traffic over the area over one hour during a day period).

Typically the area busy hour is 15–20 percent lower value than the sum of the busy hour traffics if the area contains both business and residential types. This means that a 20 percent over-capacity is needed in the radio network because of the short-term daily changes.

The long-term changes represent a clear and continuous traffic increase over an area busy hour as in the case of a new residential area or an extension of an existing area. In these cases the busy hour occurs daily at the same time and the traffic increases incrementally due to new subscribers. In another example a new office building is built in a residential area but the area busy hour is still in the evening because private subscribers are loading the radio network more than the office building subscribers during the day. In this case new transceivers are not needed for the base station because people in the office and residential buildings are not loading the radio network at the same time. Therefore, the area busy hour value is still constant and there are no long-term changes.

When the area busy hour value is measured and the number of the implemented transceivers at each base station is solved, the usage of the network (K) for a certain area can be calculated by using the equation

$$K = \frac{total\_traffic_{area\ n}}{theoretical\_traffic_{area\ n}} \qquad \text{Equation 9.1}$$

The calculated K values have to be stored in the file and in the long-term changes in these K values indicate whether the network investments are too high or low considering the radio network capacity needs. Additional KPI values like the radio network costs per air time, can thus also be easily observed.

An another cost-efficiency related KPI value is the efficient use of frequency bandwidth which is concerned with whether the radio network capacity is maximized, or in other words, C/I optimised without having reduction in speech quality. Frequency bandwidth efficiency can be evaluated by solving the frequency reuse factor that typically varies between 10–20 in the different radio networks. The actual value of the frequency reuse factor (FRF) can be easily calculated:

$$FRF = \frac{number\_of\_frequency\_channels}{average\_number\_of\_tranceivers\_per\_base\_station}$$

Equation 9.2.

The calculated value indicates whether the radio network plan is efficient. More detailed key performance indicators can also be created when an analysis of the detailed reasons of the low cost-efficiency is needed because of, for instance, inefficient use of the frequency band. Is the bad frequency efficiency either network configuration or frequency planning inaccuracy? When analysing base station coverage areas variations in base station antenna heights also have to be studied. Antenna height distribution of the base stations indicates directly whether it is possible to achieve such a low frequency reuse factors as 12–15 in a particular area even if the base station antennas are above rooftop levels.

9.1.1.2 Radio network QOS

A high radio network QOS means that *call success rate* is high while *call establishment* and a *call release* are also successful. Call establishment may be unsuccessful if radio network is congested, or if there is interference in the radio network indicating that base station congestion or blocking and interference have to be measured. In order to monitor the call release, the handover reasons, failure rates and drop call rates have to be measured. Drop call rates show where there are problems in the radio network, and handovers inform of good or poor functionality.

Call establishment

Base station radio interface blocking (lack of signalling or traffic time slots for speech or data transmission) has to be measured from the real network and compared to the Erlang formulas (Erlang-B or Erlang-C) in order to understand the traffic conditions in the radio network. The Erlang formulas work well in the normal homogenous traffic conditions but there are higher blocking values when a large group of mobile users tries to make call establishments simultaneously for instance during or after big sporting events. The analysis of base station blocking rates should be made based on the average over a day or week. The following example shows the influence of an average. There are 20 hours blocking at a base station during one week (equals a period of 7 * 24 = 168 hours) of which 18 hours have the blocking value of 1 percent and two hours have a blocking value 35 percent because of a sport event. If the average blocking over the week is calculated based on the 168 hours (including days and nights), the average blocking is

$$[148 * 0 + (18 * 1 + 2 * 35)] / 168 = 0.52 \text{ percent}$$

and if the average blocking is calculated based only on the hours when the blocking exists (20 hours) the value is

$$(18 * 1 + 2 * 35) / 20 = 4.4 \text{ percent.}$$

The values 0.52 percent and 4.4 percent are totally different and are equivalent to the case that there is the same blocking over the whole week (18 hours 1 percent and 2 hours 35 percent). This example shows that the averaging does not necessary give the best view about blocking in the radio network especially if there are high blocking peaks during the analysing period. Therefore, it is better to analyse the behaviour of the blocking e.g. to calculate the number of blockings that exceed a certain threshold value (1 percent is typically used in the GSM). This gives a better view of whether the blocking is constant or happens only occasionally due to the special conditions in the radio network service area. The value of 1.0 percent is also generally accepted as a reference for good access quality and radio network capacity should be dimensioned based on this value.

In order to analyse the behaviour of the blocking, the base station's highest traffic hours have to be found and the blocking values have to be gathered from these hours. Blocking may not be at its maximum during the maximum traffic hours because a large number of simultaneous call attempts (after events) may cause even higher values. These occasional peak values are filtered and the blocking values during the highest traffic hours correlate fairly well with the situation in the radio network. Next, the continuity of the blocking values during the day and week has to be studied and define the thresholds when the radio network capacity has to be extended. Typically the blocking threshold is 1.0 percent and is limited to exceed this threshold value for a of maximum 4–6 hours during the week (before extension) by excluding the temporary occurrences such as sport events.

The lack of signalling or traffic time slots (blocking) is the *first* possible problem area to prevent successful call establishment and thus prevent a successful call. Another reason for unsuccessful call establishment may be interference—often a reason for dropped calls during the connection. Interference is measured based on the quality classes that inform the bit-error-rates (BER) in the uplink and downlink directions. These BER values can be used as a KPI especially in the networks without frequency hopping. If frequency hopping is utilized in the radio network, the frame-erasure-rates (FER) should if possible be used or a new BER threshold is required for observation. The measured values of the interference can be averaged, for example over a week, because there should not be peaks but the interference should increase evenly with traffic increase. Typically the interference threshold in the non-hopping networks is defined to be 1.6–3.2 percent (equals the GSM quality class 4, also a quality handover threshold) to ensure a reasonable quality in the radio network. The target is to have >97 percent of connections (time or measured values) better than this GSM quality class 4. In frequency hopping networks the quality class 7 (BER >12.8 percent) has to be observed in case of BER observation and the target is to have the percentage of this quality class 7 close to zero.

Call release

The second area in the radio network QOS is the normal call release which includes the monitoring of all other types of problems which may cause an abnormal call release or dropped call in the radio interface. The *drop call rates* generally have to be monitored continuously and drop call reasons should be specified in detail. Call drops typically occur because of interference or handover failures. Interference and handover success rates have to be monitored in order to gain detail about drop call reasons. Interference is monitored in relation to successful call establishments and the same results can be used also for the drop call rate analysis. The drop call rate statistics of individual base stations require the same kind of care as measurements of radio blocking. If neighbouring base stations are blocked temporarily because of the high traffic peak, the drop call rate may also be increased if handovers to this blocked base station are prevented due to lack of traffic time slots. Hence, *the equal measurement* and *analysis strategy* has to be used for drop call rates as for blocking rates. The highest drop call rates have to be found first based on the traffic peak hours and then drop call rate behaviour has to be analysed. Finally, drop call rates of individual base stations have to be compared to the thresholds and required actions have to be commenced if the thresholds are exceeded. If more detailed studies are required the interference and handover statistics give deeper information about the situation and possible call success problem areas at an individual base station. The drop call rate threshold that triggers detailed analysis is typically 2 percent excess 4–6 times during the same week, as in the case of the radio blocking.

The handover failure or success rates and specifically handover reasons, give a detailed picture about the function of the radio network. The handover success rate is typically used as a KPI for the individual base stations to inform of probability in maintaining a call (see Table 9.1 for KPI summary). The high value (>98 percent) is expected and the more detailed information concerns the analysis of the handover reasons which also indicate whether bad functionality in the radio network arises from interference, the base station's configuration (more uplink handovers than downlink handovers → an unbalanced power budget) or something else.

Table 9.1. Key performance indicators and their recommended values for radio network monitoring in the GSM.

	Target value
Cost-efficiency	
K	0.8
Frequency reuse factor	15
Radio Quality of Service	
Blocking	1.00%
Interference	
BER (non-hopping)	> 97 % < 3.2 %
(hopping)	> 99 % < 12.8 %
Drop call rate	< 2.0 %
Handover success	>98 %

9.2 Radio network configuration

The second part of monitoring is the radio network configuration that has to be documented carefully in order to minimise errors in the radio system planning due to incorrect information regarding the radio network infrastructure. This is a valid concern when the radio network includes thousands of base station sites and one requires knowledge of an exact number of certain base station site equipment. The answer for this configuration monitoring is the database that has to include both radio network configuration and key performance indicators for the long-term monitoring (recall radio system planning documentation, Figure 2.3).

9.2.1 Databases

Radio network databases are important to have and maintain as a description of network configuration for coverage and capacity evolution; both base station configuration and base station antenna line configuration. Additionally, all parameters related to the radio network have to be documented and be available.

The *base station site database* or *base station site folder* (that was mentioned in the radio system planning process documentation) has to define the base station configuration and base station antenna line configuration.

This base station site database has to include all the necessary information about the base station equipment as

- type (indoor, outdoor)
- transmission power to the antenna port and receiving sensitivity
- combiner type
- ground height and location (*x*, *y*)
- base station antenna height, antenna directions

and all the essential information about the accessory elements (previously mentioned: low noise amplifiers, power amplifier, diversity technique, diplexers and so on). Finally, the essential radio network parameters

- frequency channels
- neighbouring base stations

have to be described in order to define the radio network functionality strategy (neighbour relations for the handovers, etc.).

This database gives the final and detailed view about the implementation of the radio network and the special characteristics of the radio planning for a certain area can be clarified. Example of the base station site folder is presented in Table 9.2

The key performance indicators (KPI) have to be stored in order to understand the evolution of radio network cost-efficiency and QOS. The area based traffic is one of the main KPI values to be monitored as it tells of the capacity changes which also effect on the QOS and thus on frequency planning or on base station configurations. These capacity changes, coverage distributions, quality issues and handovers should all be gathered in the KPI database at least every two weeks, statistically recording results that can also be analysed and mapped into the different software programs to show e.g. the traffic, blocking or dropped calls over a certain planning area.

Table 9.2. The base station site folder.

BTSID	RF_power	BTS_comb	BTS_sens	BTS_duplexer	x-coord	y-coord	top_height	ant_height
1	50W	By-passed	−106	yes	12345	123456	20	30

ant_type	ant_dir	ant_gain	ant_tilt	cable_type	cable_length	diversity	LNA_NF	LNA_GAIN
K12345	120	16	5	7/8"	50	pol.	2	14

9.3 Software programs for monitoring

The software programs are required for monitoring to visualise the obtained results from the radio network. The aim of the KPI values is to identify e.g. the number of the base stations where the blocking or drop call rate is too high. This information is not enough to show a need for a new transceiver or a new base station to solve radio interface blocking or whether optimisation work is required. Hence, monitoring results have to be presented (on a map) at exact base station locations by showing also antenna directions (and heights if required) in order to understand whether the blockings or drop call rates occur in the same location or over the network. Correspondingly, the base station configurations as base station types can be presented (on a map) to show where certain base station types are located. This information may show limitations at base stations to the use of some software features: older base stations for basic coverage usage could be located in rural areas and newer base stations for capacity usage with the latest capacity software features could be used in high capacity urban areas.

The required software programs for monitoring can be divided into three categories:

- network management system (NMS) is needed to gather the measurement results from the radio network
- A NMS post-processing tool/platform is needed to process the raw data to the required results in the tables
- A graphical presentation tool for the detailed analysis.

The NMS typically comprises radio network elements and is thus supplier related. The NMS post-processing platform can be delivered by any capable supplier who has the flexible tools to process raw data and to have a standard interface to export the results (in order to transfer the results to the graphical presentation). Finally, the results are transferred to the graphical presentation tool which has to contain a good enough digital map in order to be able to locate the problem areas. All these programs can be delivered by the same (in this case it is typically the BSS supplier) or different supplier. A good solution is to have an integrated platform (radio planning tool) which contains the interface to the NMS database (parameters and monitoring results can be sent between the NMS and the radio planning tool) and which has the capability to show the monitoring results on the digital map where the base station sites are also located.

9.4 Conclusions

Radio network monitoring is one of the key radio system planning processes as it informs of the radio network performance and monitoring results can be utilised as input data in radio network dimensioning and optimisation work. Table 9.3 gathers the key topics of monitoring.

Table 9.3. Radio network monitoring.

Subject	**Findings**
The usage of the radio network	–Has to be made over a certain area –Traffic is a key parameter –An even traffic distribution is a target
QOS	–Blocking and drop call rate have to be monitored –Averaging has to be understood exactly
Databases	–Base station site folder: · base station site configuration data · radio parameters –KPI database: · traffic, blocking, drop call rates
SW programs for monitoring	–Three level of programs are required –The final target is to visualise the monitoring results –Integrated planning platform

Chapter 10

GENERAL PACKET RADIO SYSTEM (GPRS)

10. GENERAL PACKET RADIO SYSTEM (GPRS)

The objectives of the General Packet Radio System (GPRS) are to enable the packet transmission (to have a communication "always on" and to use the radio channel, or to transmit, only when needed) and to enhance the data rate in the radio interface transmission. The GPRS is defined as a part of the GSM system and is adapted to the GSM system as depicted in Figure 10.1.

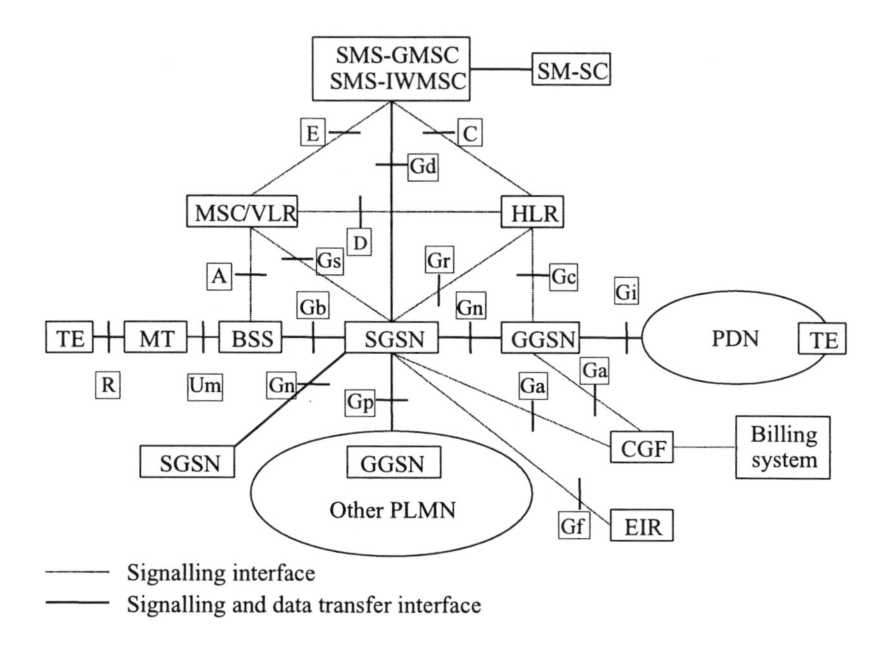

Figure 10.1. GPRS in the GSM system environment.[1]

Figure 10.1 shows that the BSS is connected to the GPRS core network via the Serving GPRS Support Node (SGSN) which is a key element when the GPRS functionality is designed and installed. The SGSN also contains functions concerning subscriber management and billing but these topics will not be discussed. Other GPRS elements or interfaces do not have a direct influence on the radio system except the Gs-interface (between MSC and SGSN) that can be used to combine the GSM and GPRS paging in order to decrease paging capacity. It was mentioned that the GPRS is needed to enable an "always on" connection which is invoiced based only on transmitted data. The other reason to have a GPRS or EDGE connection is to enhance the data transmission rates which are presented in Table 10.1.

Table 10.1. The GPRS data rates compared to the other data transmission technologies.

Technology	Schedule	Av. kbit/s	Theor. kbit/s
HSCSD	2000	58	115.2
GPRS, CS1	4Q/2000	32	72.4
GPRS, CS2	4Q/2000	48	107.2
GPRS, CS3	4Q/2001	58	124.8
GPRS, CS4	4Q/2001	80	171.2
EDGE	2001	386	500
UMTS	2002	300–400	2 000
WLAN	2000	10 000	50 000

Table 10.1 shows three different development steps in data transmission in the GSM system. High Speed Circuit Switched Data (HSCSD) represents the first step by containing the enhanced data rate up to 57.6 kbit/s (14.4 kbit/s per time slot and maximum 4 time slots for the connection). The connection in the HSCSD is still *circuit switched* and thus the GPRS is the first "always on" based data transmission. The data rate in the GPRS depends on the radio interface coding and the number of the time slots to be used. There are four different coding schemes (CS1–CS4) for the radio interface and they represent bit rates 9.05–21.4 kbit/s per time slot.[1] These bit rates depend strongly on the radio channel and they also include, e.g. headers, and thus the final application data transmission rate is smaller. The maximum number of time slots that can be used in the GPRS is defined as 8 [1] and thus the maximum theoretical data rate values 72.4–171.2 kbit/s can be determined as shown in Table 10.1. However, the average final data transmission rates in the GPRS are close to the values in the third column in Table 10.1. These values take into account the radio propagation environment and typical radio planning criteria and exclude the required information between the elements as headers. The GPRS represents the second step in the wireless mobile data transmission development and the third step contains Enhanced Data rates for GSM Evolution called EDGE and the new systems Universal Mobile Telecommunication System (UMTS) and Wireless Local Area Network (WLAN). Only the EDGE is a fully GSM

type system, the UMTS is like a next generation of the GSM (it requires a totally new radio system but it is still compatible and integrated to the GSM) and WLAN is an other standard that could be merged to the GSM in the future. The EDGE is based on the new modulation in the GSM radio interface (phase shift keying, PSK) and enables higher bit rates per time slot and while the maximum theoretical data rate is still based on transmission by utilising 8 time slots in the radio connection. UMTS has a new radio interface based on the Wideband Code Division Multiple Access (WCDMA) which enables the maximum theoretical data rate around 2 Mbit/s when the 3.840 MHz frequency spectrum is used (in the future the data rates will be many times higher when more enhanced modulation techniques are utilised). Correspondingly, the data rates in the WLAN systems are 10–30 Mbit/s when the Code Division Multiple Access (CDMA) are used in the radio interface. These "third step" systems EDGE and UMTS can be used with GPRS and thus they support the "always on" principle and the WLAN is still not compatible with the GSM/UMTS or GPRS/EDGE. The influence of these packet transmission technologies GPRS, EDGE and UMTS on the GSM radio interface and on the radio system planning is explained in this and the next chapter. First the GPRS and EDGE are explained in this chapter because they are still based on the narrowband radio propagation channel (the frequency channel bandwidth is 200 kHz in the GSM/GPRS/EDGE systems) and time division multiple access (TDMA). The UMTS is explained separately in Chapter 11 because the radio interface is changed to the wideband code division multiple access (WCDMA) that influences significantly radio system planning.

10.1 Introduction to the GPRS radio network

The GPRS service requires some changes in the radio network elements as in the base station and in the base station controllers. These changes are mainly software related for the base stations and both software and hardware related for the base station controller and their functionalities and limitations have to be concerned when dimensioning the radio network. The radio interface system planning itself is not changing too much: the radio propagation environment (chapter 1) is exactly equivalent to the circuit switched data (speech or e.g. HSCSD) and also the radio planning process (chapter 2); even the power budget (chapter 3) is similar for the GPRS radio access planning as for the circuit switched planning. The radio planning changes happen only in the detailed planning level when the planning thresholds (coverage and capacity, chapters 4 and 6) are defined. It has to be noted that the coverage and capacity planning as well as the Quality of Service (QOS) in the GPRS are related to the signal-to-noise (*S/N*) or carrier-to-interference (*C/I*) which are starting points when the radio planning criteria is defined for the circuit switched or GPRS traffic.

These coverage and capacity planning criteria are the main issues when considering the GPRS radio planning because the radio planning process is equivalent to the normal circuit switched planning. However, it has to be mentioned that the documentation and monitoring (Chapter 9), which are important post-planning functions, have a strong effect on the radio planning quality in the long-term. In the documentation some new columns and rows have to be reserved for the time slot allocation (capacity related) and the GPRS related radio parameters as well as configurations in the base station controller have to be carefully documented. Finally, the voice related key performance indicators (KPI) are not enough for the GPRS data monitoring. The new values like throughput (kbit/s) have to be observed continuously in order to know whether the mobile users can be provided with the high quality GPRS service. As a conclusion it can be noted that coverage prediction (Chapter 5), frequency planning (Chapter 7) and optimisation (Chapter 8) are equivalent for the GSM circuit switch traffic and GPRS traffic.

10.1.1 BSS related GPRS elements

The BSS related GPRS implementation means changes in the radio network and the installation of a new element called Serving GPRS Support Node (SGSN). The base station changes are only software related for the CS1–CS2 (CS = coding scheme, different coding schemes correspond to the different level of error coding and thus also different maximum data rates as shown in Table 10.1). In order to provide the coding schemes CS3–CS4 some hardware changes are usually required in the base stations and also in the transmission (16 kbit/s transmission is no longer enough). Otherwise the GPRS does not have a significant influence on base station products, only the radio planning criteria will be changed. The implementation changes are bigger in the BSC where new hardware is required for the CS1–CS2 schemes. The new item is a Packet Control Unit (PCU) card (implemented typically in the base station or base station controller [1]) that manages the GPRS traffic.

Each PCU card has its own limitations:

- Maximum number of TRXs
- Maximum number of BTSs
- Maximum number of transmission lines towards the BSC and SGSN (Gb-interface, e.g. N * 64 kbit/s)
- Maximum number of PDP contexts
- Maximum traffic (e.g. N Mbit/s per PCU)
- Maximum number of routing areas (RA) or location areas (LA)
- Redundancy.

These limitations define how the PCUs should be used and how the GPRS should be merged to the existing GSM system. Also note that *one of the above mentioned items is typically a bottleneck for capacity extensions and also typically defines the GPRS system pricing principle.* The following example demonstrates the priority of the different BSS elements for the functionality and the costs of the GPRS system.

Example 1. PCU cards and Gb-interfaces for the GPRS solution of CS1–CS2

Configuration

- How many PCU cards have to be purchased?
- How many Gb-interface have to be purchased for each PCU?
- Can these PCU cards and Gb-interfaces used freely with each others?
- What is the maximum throughput capacity of the PCU cards compared to the maximum theoretical GPRS load of each BSC (if all time slots are used for the GPRS)?

Redundancy

- How many extra PCUs are needed for the redundancy (N + 1 or 2N)?
- How many extra Gb-interfaces are needed for the redundancy (N + 1 or 2N)?
- Does the fault or error in the transmission line effect on the PCU? Fault management?

Pricing

- Which item is a bottleneck, PCU or Gb-interface?
- Why this bottleneck exists, limitation of TRX amount, base station amount, routing areas, locations areas, throughput?
- Which redundancy is needed $N + 1$ or $2N$ (Note, $2N$ redundancy is very expensive)?

Each supplier may also have limitations in their GPRS solutions to use e.g. the capacity functionalities like

- frequency hopping concepts
- overlay–underlay concepts
- something else.

This example concerned only the PCU card and the Gb-interface in the BSC. Thus, the PCU card is very critical for the GPRS capacity and system configuration but it is not the only element which effects the GPRS capacity and upon the GPRS system costs. The BSC and PCU are connected via the Gb-interface to the SGSN.

SGSN

The SGSN can be compared to the Mobile Switching Centre (MSC) as it is like a packet traffic switching centre. The SGSN is also a part of the connection between the BSS and GPRS core network (SGSN, GGSN, etc., see Figure 10.1) and therefore an understanding of the SGSN's capacity and configuration limitations can ensure the GPRS system capacity in a radio interface. SGSN again may have different limitations:

- Maximum number of transmission lines towards each BSC
- Maximum number of PDP contexts (1 – N per subscriber)
- Maximum number of subscribers connected to the GPRS network simultaneously (not transmitting but logged in)
- Maximum throughput or traffic in each PAacket Processing Unit (PAPU),
- Maximum number of PAPU units in each SGSN
- Maximum number of routing areas (RA) or location areas (LA) and their combinations between PAPU units (as well as PCU units)
- Redundancy.

Hence, the performance of the SGSN depends on its capacity and configuration limitations. The SGSN capacity is divided into PDP context management and throughput which is *x* Mbit/s by simultaneously having *y* thousand subscribers connected to the GPRS network. The critical question is what throughput (*x* Mbit/s) is enough for what number of subscribers (*y* thousand subscribers) or vice versa. If subscribers develop a habit of opening the PDP context every morning, like with a computer (or have the PDP context continuously open 24 h, this is also a GPRS application pricing question) the SGSN capacity has to be dimensioned based on the PDP contexts. It also has to be noted that *x* Mbit/s and *y* thousand subscribers correspond to the average traffic that every subscriber may use in the GPRS network (*x* Mbit/s/*y* thousand subs = *z* kbit/s) which further corresponds to the total traffic of the busy hour when multiplied by 3600 s. The forecasts of these data figures *x, y, z* may vary significantly because subscriber behaviour in the GPRS networks is a new and unknown topic.

10.2 GPRS radio planning

10.2.1 GPRS coverage planning criteria and thresholds

The GPRS coverage planning criteria can be defined by specifying the minimum required data transmission rate and thus the minimum carrier-to-interference (*C/I*) or signal-to-noise (*S/N*) ratio requirement. The *C/I* criteria is needed for high capacity areas/locations where frequencies are reused often and thus the interference level is increased. Correspondingly, the *S/N* criteria is needed for indoor locations where interference is not limiting the data rate but the key issue is the lack of coverage (rural areas are also included to this category). If the interference level is well-known (e.g. it has been measured) and it is clearly above the noise level, it determines the planning criteria for the maximum data rate at the different locations. Thus, the GPRS coverage and data throughput rate can be fixed based on the values in Figure 10.2. These values are simulated and some measurements are still needed to ensure the correct *C/I* planning criteria for the different data rates in the different radio propagation channels in practise. In this interference limited case in Figure 10.2 the coverage is already good enough but the interference is limiting the data transmission rate: interference is already at such a high level that the radio network is not any more coverage limited.

GSM900

Type of channel	Unit	Propagation conditions				
		TU3 (no FH)	TU3 (ideal FH)	TU50 (no FH)	TU50 (ideal FH)	RA250 (no FH)
PDTCH/CS1	dB	13	9	10	9	9
PDTCH/CS2	dB	15	13	14	13	13
PDTCH/CS3	dB	16	15	16	15	16
PDTCH/CS4	dB	19	23	23	23	

GSM1800

Type of channel	Unit	Propagation conditions				
		TU3 (no FH)	TU3 (ideal FH)	TU50 (no FH)	TU50 (ideal FH)	RA250 (no FH)
PDTCH/CS1	dB	13	9	9	9	9
PDTCH/CS2	dB	15	13	13	13	13
PDTCH/CS3	dB	16	15	16	16	16
PDTCH/CS4	dB	19	23	25	25	

Figure 10.2. The *C/I* requirements for the maximum data throughput rates in different environments.[1]

The noise or coverage limited case and the *S/N* based planning criteria is more complex because the planning margins should be taken into account when defining the planning threshold for the GPRS service (coverage and data transmission rate). The planning threshold definition is based in general on the downlink direction because the radio network planning tools are typically developed for the downlink calculation as it was assumed that the radio link budget is always balanced (the uplink and downlink have the same maximum allowed path loss). The normal downlink planning threshold definition for the GSM (the GPRS has not yet been considered) is started from the sensitivity of the mobile station that is –102 dBm for the 900 MHz mobiles based on the GSM specification.[2] However, mobiles typically have much better sensitivity level and the value –104 dBm can be used. Next, the required planning margins (slow fading margin, fast fading margin, interference degrade margin, body loss, mobile antenna orientation, indoor/in-vehicle penetration loss) are added to the sensitivity level and the typical planning thresholds are presented in Table 10.2. The threshold values –96 dBm and –76 dBm correspond to the outdoor and indoor planning thresholds for voice calls in the radio networks where the location probability is 90 percent (slow fading margin = 5 dB).[3]

Table 10.2 The GSM planning thresholds for the outdoor and indoor locations.

Outdoor locations		**Indoor locations**	
	dB/dBm		**dB/dBm**
MS sensitivity	–104	MS sensitivity	–104
Interference degrade margin (inc. fading)	3	Interference degrade margin (inc. fading)	3
Slow fading margin (90%)	5	Slow fading margin (90%)	5
		Indoor penetration	20
Threshold	**–96**		**–76**

The GPRS radio planning threshold has to be defined the same way by starting from the sensitivity level which has to be considered based on the *S/N* requirements for the maximum data throughput rate in the GPRS. The *C/I* margins in Figure 10.2 can approximately be used as *S/N* margins for the noise limited situation because interference and noise are a similar type of signals (coherent signals). Thus, the *C/I* criteria in Figure 10.2 can be utilized as an *S/N* margin in the noise limited GPRS case and can be added to the sensitivity level. The sensitivity level of –104 dBm already corresponds to the *S/N* = 4–5 dB for speech. Thus, a new sensitivity level for the GPRS is

– 104	(sensitivity level for the speech)
– 5	(*S/N* margin for the speech)
+ 13	(GPRS *C/I*, CS2, Figure 10.2)
= – 96 dBm.	

The margins from Table 10.2 have to be added for this new sensitivity value –96 dBm and the new outdoor threshold for the GPRS data transmission CS2 is – 96 + 3 + 5 = – 88 dBm and for the indoor locations – 96 + 3 + 5 + 20 = – 68 dBm.

These aforementioned planning thresholds are for the noise limited situations and the definition was based on the *C/I* simulations: it was assumed that the interference and noise are similar types of signals. Hence, the *S/N* based measurements have to be performed to confirm this theoretical determination and to depict Figure 10.2 again by using the correct *S/N* instead of *C/I*. The value 13 for the *C/I* or *S/N* is very important because it almost provides the maximum transmission data rate when the CS2 coding scheme is used and because the value 13-15 is very widely used as a frequency planning criteria for the voice calls in the GSM.

The equal planning thresholds can also be determined for the interference limited case. The average interference level has to be measured in the radio network and the planning margin in Figure 10.2 has to be added to the average interference level in order to ensure the required data transmission rate. If the interference level is for example –110 dBm inside the building, 13 dB has to be added in order to provide the maximum data rate of the coding scheme 2. Thus, a radio planning threshold is –110 dBm + 13 dB = –97 dBm (GPRS sensitivity level) + 3 + 5 = –88 dBm inside the building. This example shows also that the interference level of –110 dBm and the mobile station sensitivity level of –104 dBm require about the same radio planning threshold.

10.2.2 GPRS capacity planning

The GPRS capacity planning in the base station subsystem (BSS) includes the radio interface and base station controller capacity planning. Both radio interface and BSC capacity planning have to take into account that the GPRS data transmission is a new form of traffic in the GSM radio network. The GPRS traffic is based on packets and thus its duration is mostly very short compared to the circuit switched traffic as speech. The conversation on average takes from 60 s to 120 s (1–2 minutes, sometimes of course only 15 s and sometimes 1 hour) and Erlang-B formulas and tables can be used to define the required capacity for the radio network. The GPRS traffic may also be almost like speech in duration when a large data amount is transmitted e.g. a transmission of 2 Mbit takes 40 s if the transmission rate is 50 kbit/s. However, this 2 Mbit transmission is quite rare and it takes only 40 s (<60–120 s). Typical data transmission may be around 300 kbit which takes only 6 s time to transmit. This 6 s is quite a short-term traffic and thus Erlang-B formulas may not be valid any more.

The final capacity of the GPRS depends on the radio network configuration and on the capacities of the GPRS related elements as the

- number of time slots in the radio interface Um
- PCU card capacity in the BSC

- number of the Gb-interfaces between the BSC and SGSN
- processing power of the PAPU units in the SGSN.

If any of these capacities is a bottleneck, the service level (data rate) corresponds to the maximum data rate of the bottleneck unit and the QOS may deteriorate (data rate reduction). The capacity requirements of the PAPU units in the SGSN and PCU cards in the BSC are easy to define because the traffic from the various subunits (the traffic from the several base stations to the PCU card and the traffic from the several BSCs to the PAPU unit) can easily be monitored and divided between the PAPU and PCU elements. Thus, the capacity reservations can be done efficiently based on the observations from the network and by following the average traffic increase. Thus, the major GPRS capacity problem is in the radio network because the maximum capacity reservation of 4–8 time slots corresponds to the capacity over 15–30 percent of the total capacity of a typical 3 transceiver base station.

Radio interface capacity planning

The radio interface for the GPRS has to be dimensioned based on a specific traffic model as mentioned. This traffic model can finally be verified when the GPRS traffic is monitored. The capacity management can be made easier by having the dedicated circuit switched and GPRS time slots and a group of time slots for both usage dynamically based on the traffic requirements. The circuit switched traffic typically always has a higher priority to use these "hybrid" circuit switched and GPRS time slots. The "optimistic" way of calculating the maximum available GPRS traffic at the base station is based on the circuit switched traffic and Erlang-B formulas presented in the following example.

Example 2. GPRS traffic in Erlangs

2 TRX/base station = 8.1 Erl at 1 percent blocking
(16 time slots = 1 signalling + 15 traffic)
15 traffic channels continuously used = 15 Erl
⇒ 15 Erl – 8.1 Erl = 6.9 Erl for the GPRS

This example is optimistic because it assumes that the GPRS traffic demand varies and always corresponds to the traffic that all the available time slots (not used for the circuit switched traffic) can offer: based on this principle the 2 TRX capacity would be 15 Erl and there is no blocking of GPRS calls at all. This example shows an extra capacity that can be used for the data transmission if the data throughput rate is not concerned. Thus, the available traffic (available number of time slots) for the GPRS varies in this example from zero to N (N is maximum that can be used) and the minimum throughput that is needed for a certain service like for the video transmission can not be guaranteed; a good point when concerning telematic applications (watch dogs) for which it is enough to send the message for example during the next hour. For these cases the example can be used because these telematic messages always can be sent when there is a space in the radio interface. As a conclusion it can be noted that the GPRS traffic includes applications for which the transmission rates have to be guaranteed and applications that can be sent when possible and the maximum traffic over the radio interface is something between 8.1 and 15 Erl in the example.

Moreover, the capacity of the base station can be analysed by defining each transceiver (8 time slots = the maximum data rate = 96 kbit/s, CS2) or smaller set of time slots for the GPRS transmission as an independent transmission channel. Thus, there would only be 3 "transmission channels" in the base station in the case of 8 time slot transmission and 3 transceivers, and the capacity "could be estimated" by using Erlang-B tables.

Example 3. The capacity of the 96 kbit/s data rate in the 3 transceiver (TRX) base station

3 TRXs = 3 * 96 kbit/s channels = 0.46 Erl at 1 percent blocking
average data transmission 300 kbit → 300 kbit / 96 kbit/s
= 3 s = 0.8 mErl
number of the served subscribers per busy hour = 0.46 Erl / 0.8 mErl
= 575 Subs.

The same 3 TRX capacity can serve over 900 speech subscribers when each subscriber loads the network 15 mErl. Moreover, first note that the figures 575 and 900 depend fully on the traffic per subscriber. Second, note that the figure 0.46 Erl is based on the Erlang-B formula that may not be accurate for the GPRS traffic. Additionally 1 percent blocking is too tight a criteria for the data traffic, which could be for example 10 percent.

PCU, Gb, SGSN capacity planning

The radio interface is the first bottleneck in the BSS when dimensioning the GPRS capacity. The second bottleneck is the PCU card that is typically implemented in the BSC as shown in Figure 10.3. The PCU card throughput is an essential value that has to be compared to the maximum number of transceivers and base stations that can be connected to each PCU card. Moreover, the PCU card is connected to the SGSN (if the PCU cards are implemented in the BSC) via the Gb-interfaces that carry on the data and signalling.

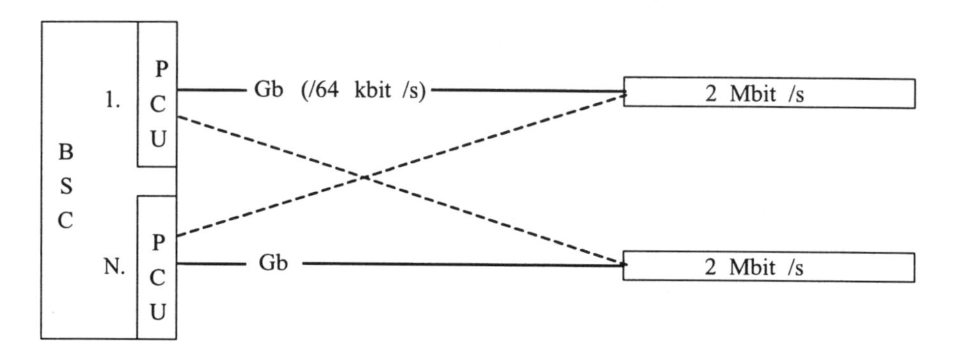

Figure 10.3 BSC capacity dimensioning for the GPRS traffic.

The major capacity planning activity is to understand the traffic that is coming from the radio network via the PCU cards. If the traffic from the radio network is constant (especially the maximum traffic) the base stations can be divided into different PCUs and the required number of 64 kbit/s Gb-interfaces can be allocated. The capacity planning rule for the PCU is that as few PCUs should be used because the trunking effect is the maximum when the resource pool (Gb-interfaces) is maximised: it is better to have 8 Gb-interfaces in the 2 PCU cards than to have 2 Gb-interfaces in the 8 PCU cards. The negative point of this configuration is that one day PCU cards are overloaded and have to be reconfigured and the base stations have to be moved to new PCU cards and changes may also be done in the SGSN. If the traffic is not constant it is not easy to divide the base stations to the different PCU cards. Additionally, the redundancy in the Gb-interface has to be done based on the $N + 1$ or $2N$ principle. The $2N$ requires a lot of extra Gb-interfaces and thus the 64 kbit/s links are easily limited. Finally, recall the maximum capacity of the SGSN that includes the Mbit/s and number of PDP contexts. It can also be concluded that the PCU/Gb capacity management is dependent on the GPRS monitoring.

10.3 GPRS related radio network functionalities

The GPRS contains some new and modified functionalities in the radio interface compared to the basic GSM specification. These functionalities are related to the

- location/routing area (LA/RA) and paging
- mobile management (MM)
- radio resource management (RR)
- base station re-selection (no handover, network controlled handover is an option in the GPRS, *mobile is master*).

Location/routing areas (LA/RA) and paging

The GPRS signalling principles are equal to the circuit switched signalling principles except the paging that is based on routing areas (RA). The routing area is a part of the location area or equals the location area but it is never larger than the location area. The routing area is the most exact unit that refers to the mobile station location when there is no data transfer ongoing between the mobile station and base station. All the base stations are always paged inside the routing area when a new GPRS data transmission is activated from the network. Paging can be done by using the circuit switching paging or by using the separate GPRS paging. The circuit switching paging is enough (the separate GPRS paging is not needed) if the location areas are small enough: the maximum number of pages per busy hour that can be transmitted over the radio interface is around 50000 when only one time slot and two CCCH blocks are used for the signalling and paging in any of the base stations inside the location area (the worst case). Thus, paging should not be an issue at the early stages (during the first 2–4 years) of the GPRS.

Mobility management

Mobility management means that the mobile station location has to be known with a certain accuracy and includes all the different GPRS serving states. The mobility management states (in the GSM IDLE and DEDICATED) in the GPRS are divided into IDLE, STANDBY and READY.[1] IDLE means that the mobile station is not connected to the GPRS core network (there is no PDP context activation).

Correspondingly, STANDBY means that the mobile station is connected to the GPRS core network (PDP context activation has been done) and READY means that there is a request (mobile originated or mobile terminated) to transmit data over the radio interface. READY therefore refers to DEDICATED and STANDBY to IDLE in the GSM.

In IDLE mode the GPRS network does not know anything about the mobile, and in the STANDBY mode the mobile station's location is known at the routing area level. The state is changed from STANDBY to READY by sending a page from the network—only mobile terminated GPRS calls—or when the mobile sends a service request in the uplink direction. In READY mode the GPRS network knows the mobile station location at the base station level.

Mobility management in the GPRS also includes the base station re-selection and location update procedures especially at the routing area boundaries. The location update has to be done every time the mobile station changes routing area. The routing area boundary is critical and some hysteresis (the same parameter cell_selection_hysteresis as in the GSM) is needed in order to prevent continuous location updates if the mobile station is only in the centre of the base station coverage areas.

Radio resource management

Good radio quality can only be ensured by guaranteeing enough radio capacity without interference (radio resource management). This radio capacity includes both the traffic channels and signalling channels. The traffic channels are typically divided into the "only for the GPRS users = *GPRS*" and "for both circuit switched and GPRS users = hybrid *GSM/GPRS*" time slots. Additionally, the time slots on the BCCH transceiver can have for example higher priority for the GPRS usage or vice versa and in advanced solutions the GPRS capacity can also be reserved from different frequency layers (900/1800 MHz). Besides these freedoms there may also be some limitations in the radio interface due to the capacity features like overlay–underlay concepts or frequency hopping as already mentioned before.

Signalling happens in the GPRS almost like in the GSM and there are no extra requirements for the GPRS. The required GPRS paging capacity amount is often discussed but recall that paging happens only when there is a mobile terminated (network originated) call and the page is sent only once when the data transmission starts (the mobile station changes the state STANDBY→READY). This paging is not a problem if the location area planning has been done in a sensible way *even if the routing areas are set equivalent to the location areas*.

The SDCCH or other signalling channels are not loaded more than in the GSM either, actually e.g. the handover commands are left out from the first releases of the GPRS because there are no network controlled handovers.

Base station re-selection

In order to ensure mobility and good quality (high enough data rate) in the GPRS the data transmission has to continue even if the mobile station is handed over to the next base station. This is provided by the base station re-selection based on the normal C1 or C2 criteria or on optional GPRS criteria C31 and C32.[4] The mobile station measures the serving base station and the neighbour base stations every 3–5 s both in the "READY" (GPRS call is on) and "STANDBY" (no GPRS call but the PDP context is activated) modes and makes the decision on the new serving base station based on these measurement results.[3] Thus, the base station re-selection may happen every 5 s even if the neighbour base station is only 1 dB better if no hysteresis is used. The hysteresis (= margin, parameter cell_selection_hysteresis [4]) for the re-selection is used only at the location and routing area borders and only in the "IDLE" mode in the GPRS. A new parameter called ready_state_cell_selection _hysteresis [4] is required to define the base station's re-selection hysteresis between the base stations *within the same routing area in the GPRS "READY" mode*.

10.4 GPRS parameters

The GPRS related radio parameters can be divided based on the GSM categories *signalling*, *radio resource management*, *mobility management*, *measurements* and *handovers* and *power control*. It can be noted that signalling has to be emphasised because the separate packet transfer signalling channels are introduced in the GPRS [5] and because the GPRS cause an extra load on the signalling. Radio resource management also needs some new radio parameters in order to handle the time slot reservations for the GPRS transmission. Moreover, the mobility management and measurements are not changing at all, nor the handovers and power control only slightly. Hence, the GPRS does not cause too much extra work in parameter planning and it is enough to take care the location and routing area sizes (the easiest way is to have the location areas equivalent to the routing areas) in order to ensure enough paging capacity in the signalling and then to make a good strategy for the radio resource management. In the radio resource management it has to be decided whether the GPRS time slots are on the BCCH frequency or on the second or third transceiver or whether the 900 or 1800 MHz frequency channels are prioritised. Almost all these radio resource management related parameters are supplier related. After defining the signalling and radio resource management almost nothing can be done by the network in order to control the GPRS call because the mobile station decides the cell re-selection by using the hysteresis parameters in the different operating modes.

10.5 GPRS monitoring

The basic GSM monitoring, as explained in Chapter 9, can be used as a basis for the GPRS monitoring as well. The traffic and blocking KPI values give a direct and rough indication about the situation at each base station when concerning the capacity. The basic GSM counters can also be used efficiently for the evaluation of the GPRS transmission quality, like throughput rate at each base station. For example the received power level counters in the NMS can be set to measure the percentage of the received power level values that exceed the threshold that corresponds to the maximum throughput rate in the GPRS when using the different coding schemes. The power level threshold for the maximum throughput rate can be calculated from the base station's and mobile station's sensitivity values as the planning thresholds. When these power level measurements are performed from the operative network they represent the actual situation in the radio network and thus these results can be used in confidence when evaluating the GPRS coverage. However, some counters are needed for the GPRS specific monitoring in order to specify the overall GPRS transmission quality.

The GPRS monitoring is important because the planning thresholds especially for the capacity are still more or less open. It also has to be remembered that the whole monitoring is related to the capacity or data rate observations because all the issues—as lack of coverage or lack of time slots—have an influence on the data rate: there is actually no such hard blocking directly in the GPRS as in the GSM when speech is considered. Thus, the critical monitoring item is a throughput or kbit/s. The deterioration of the data throughput always indicates that there is a coverage or interference problem in the radio network. The detailed measurements and analysis are always needed to solve the reason for this overall quality reduction.

10.6 Enhanced Data rates for GSM Evolution (EDGE)

Enhanced Data rates for the GSM Evolution (EDGE) is an improved packet data transmission technology in the GSM based on the radio channel using 200 kHz bandwidth. EDGE is based on the phase shift keying (PSK) modulation which makes it possible to increase the data throughput per channel higher than 60 kbit/s (see Table 10.1 and thus the maximum data throughput is more than 500 kbit/s when eight time slots are used for the transmission.

The GPRS was a new type of technology because the traffic is a packet type (no longer circuit switched), there are new elements required to handle the packet type of traffic and the functionality of the radio network was also changed due to mobile station independence (the mobile station is not a slave in the GPRS because it controls the base station re-selection procedure). The radio system planning process in the GPRS is similar to that of traditional GSM and only some changes are needed in the detailed level planning for the coverage (new planning thresholds for the noise and interference limited cases) and capacity issues (new traffic model). The EDGE does not change anything compared to the GPRS radio planning except new planning thresholds are required again for the coverage planning. The traffic is still packet based and thus the same traffic model (maybe with some slight modification) can still be used in the EDGE as in the GPRS. Hence, the EDGE does not have a significant effect on radio system planning in the detailed level but for the long-term strategy it has a strong influence because the base station locations should be optimally selected for the GSM, GPRS and EDGE services which have in the worst case almost 20 dB difference in coverage planning thresholds. This long-term strategy becomes still more complicated when the wideband UMTS system at 2100 MHz frequency band is taken into account in the radio planning process.

10.7 Conclusions

At a general level it can be concluded that the GPRS or EDGE radio planning is quite equivalent to the GSM and the GPRS/EDGE planning challenges arise from capacity planning (and from the coverage planning for the indoor and rural areas) (see Table 10.3). The GPRS or EDGE traffic demand (traffic model) from the radio network is a critical parameter that has to be optimised in order to maximise the radio Quality of Service (the coverage, capacity, interference level) and cost efficiency. Additionally, radio network monitoring has to be emphasised in order to get the information about the behaviour of mobile users.

Table 10.3 GPRS radio system planning.

Subject	**Findings**
GPRS radio propagation channel and environment	–The GPRS as well as the GSM have the same radio propagation principles –The GPRS and GSM radio system planning are equivalent
PCU, Gb, SGSN network elements	–The PCU, Gb, SGSN units are critical for the configurations and capacity
Coverage planning	–Interference or noise limited –Power budget equals to that of GSM –Planning margins equivalent –New coverage threshold
Capacity planning	–Traffic model still unclear –Capacity planning based on traffic monitoring
Monitoring	–Monitoring is critical –kbit/s should be observed –The received power level statistics indicates the GPRS coverage area
EDGE	–8-PSK modulation → higher data rate –New coverage thresholds and base station sites –The EDGE radio system planning equals to that of GPRS and GSM

10.8 References

[1] ETSI, Digital cellular telecommunications system (Phase 2+), General Packet Radio Service (GPRS), GSM 03.60.

[2] ETSI, Digital cellular telecommunications system (Phase 2+), Radio transmission and reception, GSM 05.05.

[3] W.C. Jakes, Jr., (ed.), "Microwave Mobile Communications," Wiley-Interscience, 1974.

[4] ETSI, Digital cellular telecommunications system (Phase 2+), Function related to Mobile Station (MS) in idle mode and group receive mode, GSM 03.22.

[5] ETSI, Digital cellular telecommunications system (Phase 2+), Overall description of the General Packet Radio Service (GPRS) Radio interface; Stage 2, GSM 03.64.

[6] ETSI, Digital cellular telecommunications system (Phase 2+), General Packet Radio Service (GPRS); Mobile Station (MS) – Base Station System (BSS) interface; Radio Link Control / Medium Access Control (RLC/MAC) protocol, GSM 04.60.

[7] ETSI, Digital cellular telecommunications system (Phase 2+), Radio subsystem link control, GSM 05.08.

Chapter 11

UNIVERSAL MOBILE TELECOMMUNICATION SYSTEM (UMTS)

UNIVERSAL MOBILE TELECOMMUNICATION SYSTEM (UMTS)

11.1 Introduction to UMTS

The objective of the Universal Mobile Telecommunication System (UMTS) is to provide wireless services, which require higher *data* rates than can be offered in the GSM/GPRS/EDGE radio networks. The UMTS is thus an evolution for the GSM because it can offer the voice calls but also much higher bit rates for the wireless mobile connections. Hence, the critical improvement in the UMTS is data transmission enhancement compared to the GSM radio interface. In the future the UMTS may be used more for data transmission than for voice calls and the GSM will be used more for speech because it already has good enough voice quality. Because data transmission is the key issue in UMTS the discussion turns to the data rates that can be achieved in the UMTS and to the services that can be provided for mobile users.

These mobile data *services* can be divided into different classes based on content (banking, entertainment, etc.) or based on the required "media" e.g. short messages, pictures or video channel. These different "media" reserve the radio network in a different way (one short message reserves only few a percent of the base station total capacity and a video channel may reserve 10–50 percent of the capacity) and thus the radio network dimensioning has to be done based on these service and capacity requirements.

The key issue is to understand the capacity requirements of the different services and to adapt these services and capacity needs in the radio networks. Thus, the conversation between the service providers and the radio network operators has to be continuous and done well in order to agree the services that can be delivered over the whole or partial radio network. This discussion needs careful business plans, which again need an understanding of the service portfolios and also of the radio network *implementations*. The key instruments in the UMTS evolution—data, services and implementation—and the GSM/UMTS integration are seen in Figure 11.1.

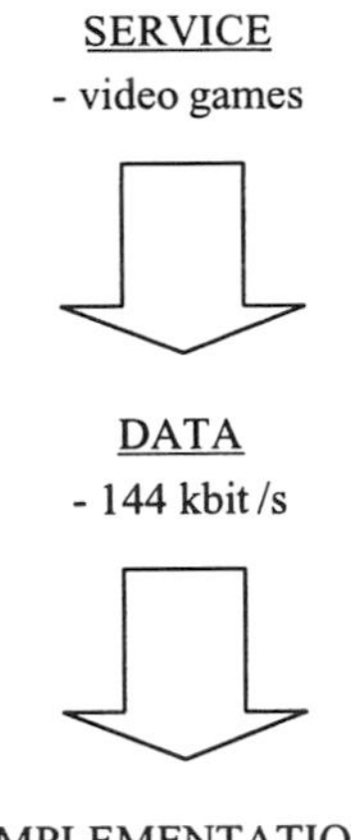

Figure 11.1. The UMTS key instruments in the GSM/UMTS integration.

The implementation and implementation strategy have to be planned carefully because the UMTS is a totally new system. The UMTS operates in the frequency band of 2100 MHz which is more than double compared to the 900 MHz and clearly higher than 1800 MHz which are typically used in the GSM (1900 MHz in the USA). These differences in the operating frequencies mean that the radio propagation is not equivalent and the old base station coverage areas are not necessarily valid in the UMTS frequency band. This generates problems in implementation because the old base station sites would be very cost-efficient to reuse but simultaneously they are necessarily not the most optimal locations for the coverage. This selection of the base station site reuse depends heavily on the implementation strategy (a continuous or discontinuous UMTS coverage) and on the traffic forecasts: whether the continuous UMTS radio network will be implemented within 3 years or in 10 years period.

WCDMA (Wideband Code Division Multiple Access) was selected as the radio interface technology of the UMTS systems. It is totally different from the technology used in the GSM. Fortunately the basic radio system planning philosophy does not change but almost all the detailed planning items concerned, e.g. the power budget, have to be checked and adjusted to be suited for the WCDMA technology. Also the radio system planning process has to be modified slightly from the traditional model because the traffic can vary from the speech to the 2 Mbit/s data and it can be either circuit switched or packet based. All these detailed differences between the GSM and UMTS radio system planning will be covered in this chapter simultaneously explaining the UMTS radio system planning as an evolutionary step to the GSM radio system planning.

11.2 UMTS radio interface

11.2.1 WCDMA air interface specification

The WCDMA specification has certain key features that are listed below:[1–5]

- DS-CDMA
- Chip rate 3.84 Mchip/s
- FDD and TDD modes
- Channel bandwidth about 5 MHz with center frequency raster of 200 kHz
- Multirate and -service
- 10 ms frame with 15 time slots.

WCDMA air interface is based on direct sequence code division multiple access (DS-CDMA) technology. This means that the user data sequence is multiplied with a so-called spreading sequence (DS), whose symbol (also called chip) rate is much higher than the user data rate; this spreads the user data signal to a wider frequency band. The relation between user data rate and chip rate is called a spreading factor ($spf = R_{chip}/R_{bit}$). The chip rate in WCDMA is 3.84 Mchip/s and spreading factors are in the range of 4 to 512, thus the user net bit rates supported by one code channel are in the range of 1 to 936 kbit/s in the downlink. Up to 3 parallel codes can be used for one user, giving bit rates up to 2.3 Mbit/s. In the uplink data rates are half of these figures, because of modulation differences.

WCDMA standard includes two modes of operation, WCDMA/FDD and WCDMA/TDD. WCDMA/FDD is a frequency duplex mode, in which the uplink and downlink signals are at different frequency bands. WCDMA/TDD is a time domain duplex mode, where the uplink and downlink signals are at the same frequency but separated to different time periods. *Later in this book WCDMA refers to WCDMA/FDD* because this will most probably be the air interface, which is first deployed and utilised.

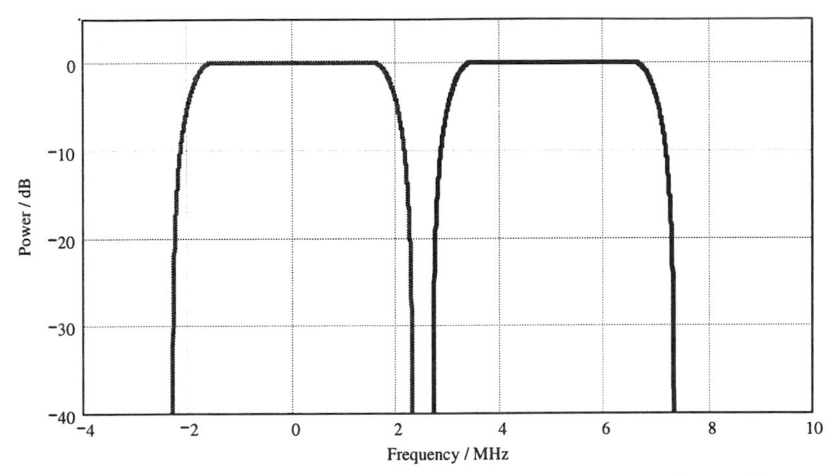

Figure 11.2. Theoretical spectrum of two WCDMA carriers with 5 MHz channel spacing.

The nominal channel bandwidth of WCDMA signal is 5 MHz as illustrated in Figure 11.2. The specification gives the flexibility to define the exact channel centre frequency at 200 kHz raster, so that the actual channel separation utilised might be smaller than the nominal 5 MHz down to the minimum of 4.4 MHz specified. This has to be noted carefully because it might cause interference in the network.

The aim of UMTS radio interface specification and selection work was to achieve a system which has the flexibility to support a large range of different services, some of them not even thought of today. This is implemented as a support for largely different data rates and different channel types from circuit switched to shared packet channels. The WCDMA radio interface also supports multiple simultaneous services via different channel types offered to one user.[6]

The transmission in WCDMA is split to 10 ms radio frames each of which consists of 15 pieces of 666 μs (2560 chips) timeslots. The bit rate and, for example, channel coding can be changed every 10 ms frame, offering very flexible control of the user data rate. Every timeslot has bits reserved for pilot signal, power control (TPC bits), transport format indication (TFCI bits) and also if needed closed loop transmit diversity (FBI bits). The exact signal format and multiplexing is quite different in uplink and downlink signalling.[6,7] Also the dedicated and shared channels have several differences in signal format.

11.2.2 Propagation environment

The radio propagation environment was divided in Chapter 1 into outdoor and indoor classes and the outdoor class further divided into macrocellular and microcellular propagation environments and moreover the macrocellular type of environment can still contain different building densities such as urban, suburban or rural type of area. The grouping of these different classes is depicted in Figure 11.3. Each of these propagation environments has special characteristics of the radio propagation channel.

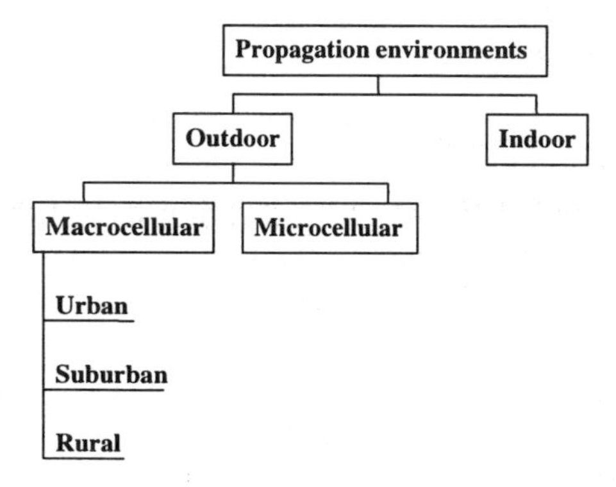

Figure 11.3. Radio propagation environment classes.

The radio propagation channel typical to each radio propagation environment class can be characterised by the following main properties

- angular spread
- delay spread
- fast and slow fading characteristics
- propagation slope.

When we want to understand the difference between GSM and UMTS radio interface performance, the most important property of the channel is the delay spread. That describes the amount of multipath propagation in the propagation environment of the radio link. The delay spread can be calculated from the typical (estimated or measured) power delay profile, which describes the signal power as a function of the delay. Power delay profile can be presented also as impulse (power) response of the channel. Figure 11.4 presents an example of power delay profile based on the channel model defined in reference [8].

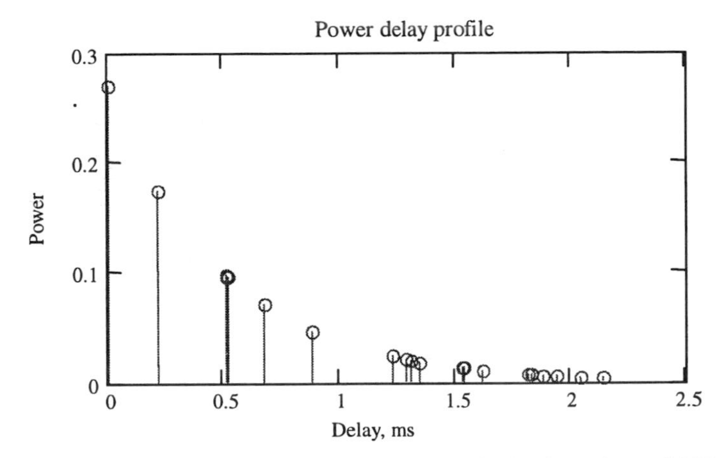

Figure 11.4. Channel impulse response, typical urban channel.[8]

The delay spread $S\tau$ can be calculated from the power delay profile, as was presented in Chapter 1. Typical values of delay spread for different environments are presented in Table 11.1.

Power delay profile and delay spread are time domain properties of the radio channel. The effect of the multipath to the radio channel can also be described by the frequency domain properties of the radio channel. In the frequency domain multipath causes frequency selective fading, signals at different frequencies have different fading (amplitude and phase). The frequency response of the channel can be calculated as fast fourier transformation (FFT) of complex impulse response of the channel. Figure 11.5 presents the frequency response of the channel, whose power delay profile was presented in Figure 11.4.

One frequency domain property of the channel is coherence bandwidth Δf_C. It can be calculated from the time domain property delay spread as was shown in Equation 1.3. Coherence bandwidth is the minimum frequency separation of two carriers, which have significantly uncorrelated fading.

Table 11.1 shows the calculated coherence bandwidths typical for different radio propagation environments.

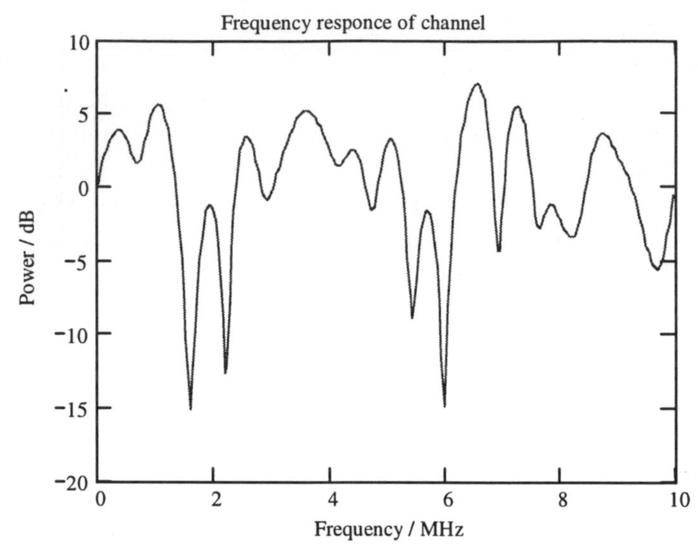

Figure 11.5. Frequency response of multipath channel presented in Figure 11.4.

Often when different air interface technologies like WCDMA and GSM are compared one argument is whether the system is narrowband or wideband, WCDMA is called wideband and GSM narrowband systems for example. The system is narrowband when the radio signal bandwidth is much smaller than the coherence bandwidth of the radio channel and wideband when it is much larger. Thus that system property depends strongly on the typical propagation environment in which the system is used and might be different in different environments. This is shown in Table 11.1. We can see, for example, that in typical urban channel UMTS is a wideband system and in the GSM neither clearly narrow- or wideband. But in the typical indoor cell environment both of these systems are narrowband systems.

Table 11.1. Characteristics for different radio propagation environments.[8]

	Delay spread, μs	Δf_C, MHz	WCDMA	GSM	IS-95
Bandwidth, MHz			3.84 MHz	0.27 MHz	1 MHz
Macrocellular					
Urban	0.5	0.32	WB	NB/WB	WB
Rural	0.1	1.6	NB/WB	NB	NB
Hilly	3	0.053	WB	WB	WB
Microcellular	< 0.1	> 1.6	NB/WB	NB	NB/WB
Indoor	< 0.01	> 16	NB	NB	NB

WB = wideband, NB = narrowband

As was mentioned the coherence bandwidth is related to the correlation of fading between different frequencies in a channel. In a propagation environment where a system is narrowband the fading is frequency non-selective or also called flat. In the wideband environment fading for signal frequencies is uncorrelated and the fading is called frequency selective. Figure 11.7 shows the UMTS and GSM system signals in a typical urban [8] channel. We can see that the fading is clearly frequency selective for the UMTS signal but for GSM the fading is more flat.

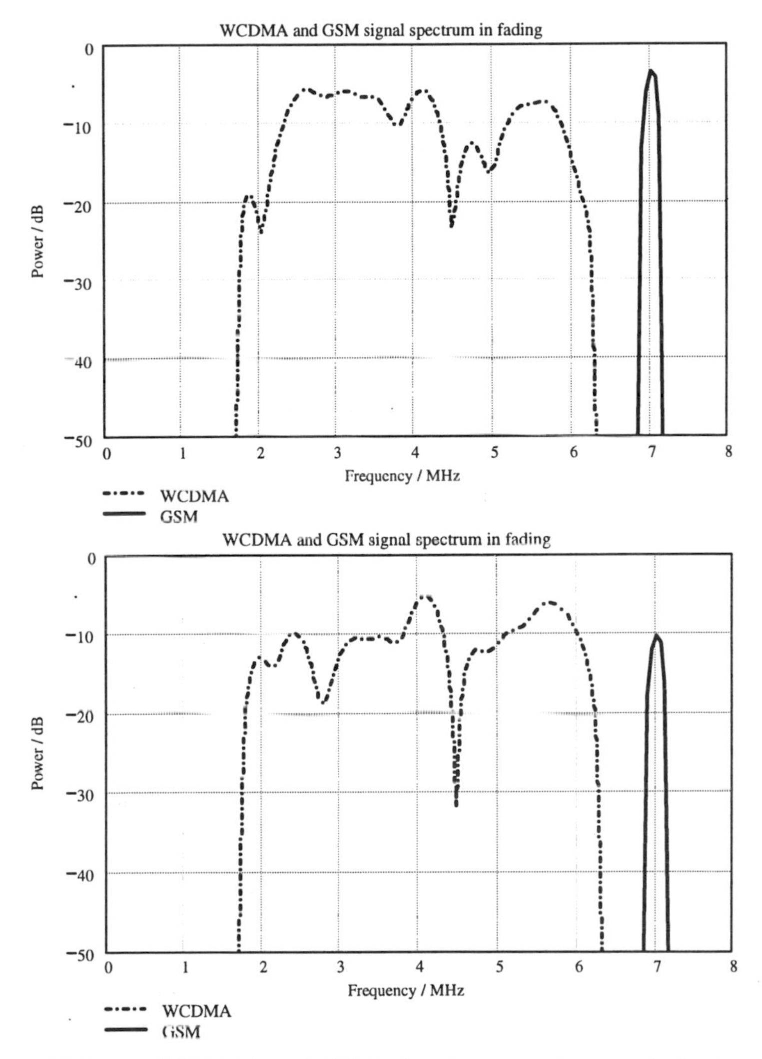

Figure 11.7. WCDMA and GSM signal spectra in a typical urban fading channel at different positions d_1 and d_2.

As explained in Chapter 1, slow fading is caused by buildings or other structures and natural obstacles attenuating the radio signal. This variation of signal level does not directly depend on the system used (for example GSM or UMTS); the only factor, which has to be taken into account, is the signal carrier frequency. The slow fading characteristics of the radio channel are different for each signal at different frequencies, but how exactly they are changed is not well known. Implicitly one could conclude that the standard deviation of the signal level is higher for higher frequencies, because the signal is affected more by obstructions.

The same logic that was used for slow fading characteristics applies also to propagation slope. Thus in practical planning the best way to take into account possible changes in propagation slope characteristics is to tune the propagation model that is used.

11.2.3 Receiver performance

The main component of a CDMA receiver is a so called correlator, which consists of code multiplication and integration units. A code generator is also required to generate the local replica of the spreading code used to wipe off the spreading code from the received signal, to de-spread the signal. This is represented in Figure 11.8.

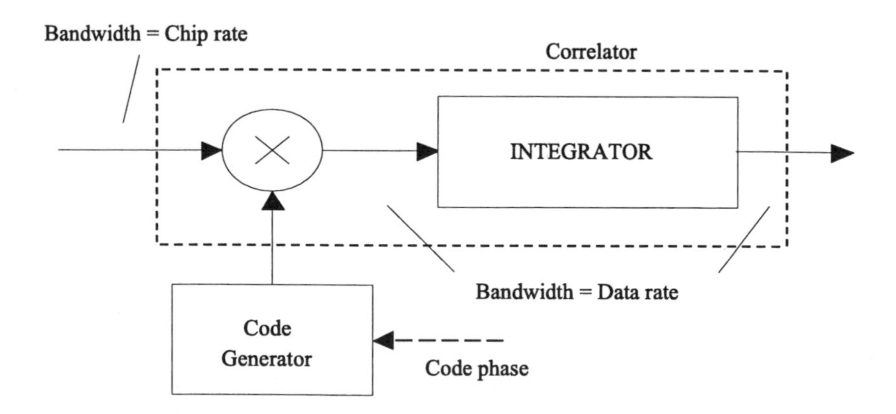

Figure 11.8. CDMA receiver correlator and code generator.

The spreading code autocorrelation function has a high peak only at zero code phase difference (Figure 11.9). At larger than 1 chip code phase difference ($\Delta\tau$) the signal will be attenuated about factor $1/spf$ (spf =

spreading factor). The exact attenuation depends on the exact autocorrelation function of the spreading code.

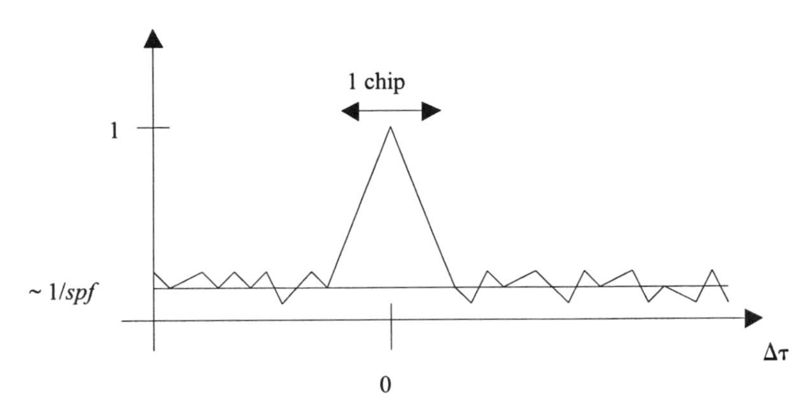

Figure 11.9. The autocorrelation function of a spreading code.

Because of the autocorrelation function the code phases of the incoming signal and local code replica has to match within a fraction of chip time in order to maximise the received signal level at the output of the correlator. The chip duration of WCDMA signal is

$$T_{chip} = \frac{1}{R_{chip}} = \frac{1}{3.84MHz} = 0.26\mu s \qquad \textit{Equation 11.1}$$

When we compare the chip duration to the typical urban channel multipath impulse response, for example the one presented in Figure 11.5, we can see that the maximum delay difference between the first multipath component and the last one (~ 2.2 μs) can be much larger than the WCDMA chip duration (0.26 μs). One correlator receiver is capable of receiving the signal power falling within the window of roughly one chip, the rest of the signal is attenuated by the spreading factor (Figure 11.9). Optimum performance is achieved when all the multipath components can be used. This has led to the use of a Rake receiver structure utilising several parallel correlators (also called fingers). Figure 11.10 shows the main structure and blocks of a Rake receiver.

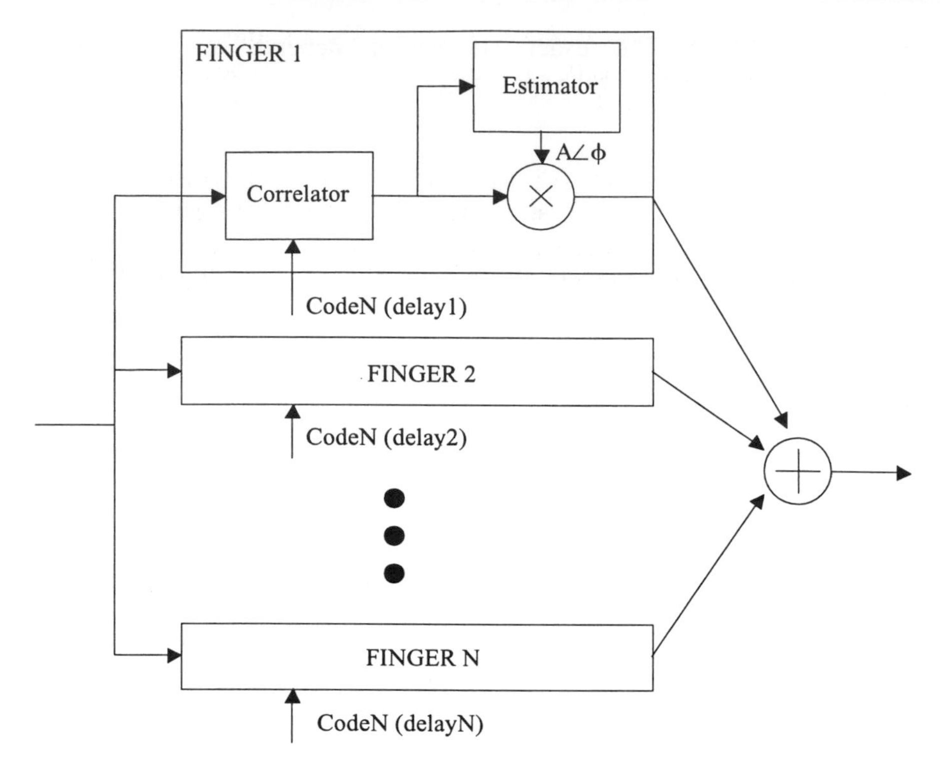

Figure 11.10. Rake receiver block diagram.

In addition to the correlator, each Rake finger has a phase and amplitude estimator. The phase estimate is used to equalise the phases of the fingers before the summation and the amplitude estimate is used to weight the fingers with their respective amplitudes. This combining method is called Maximal Ratio Combining (MRC).[9] It is proved to be the optimum method to combine several signals. One basic property of MRC is that the fingers with no signal are omitted from the summation to prevent the increase of noise level.

A channel impulse response estimator is required to allocate the delays for the Rake fingers. The estimation and allocation process is continuous as the channel changes continuously when the receiver moves.

The performance of a Rake receiver depends on the radio propagation channel and its multipath propagation conditions. As mentioned before the aim of a Rake receiver is to receive and combine the relevant channel multipath components. The larger the number of multipath components that can be separated (delay difference is larger than chip time) the better is the performance of the Rake receiver. That number of multipath components can also be called multipath diversity as each finger can be seen as a diversity channel.

In Figure 11.11 the impulse response of a typical urban channel [8] is presented along with one possible allocation of Rake finger delays and amplitudes. It can be seen that a significant number of Rake fingers can be allocated in this channel condition. Thus the amount of multipath diversity is high and the performance of the Rake receiver is good.

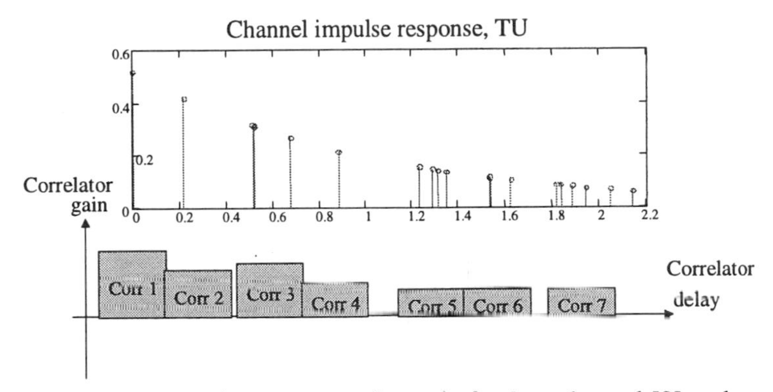

Figure 11.11. Impulse response of a typical urban channel [8] and example of Rake receiver finger delay and amplitude allocation for this channel condition.

In Figure 11.12 an example of an indoor channel impulse response is shown. All the significant multipath components fall within one chip interval and the Rake receiver cannot separate them. In conditions like this the fading is flat and the system is narrowband (Table 11.1), there is no multipath diversity. The chip rate and signal bandwidth should be much larger in order to separate the multipath components and to utilise multipath diversity.

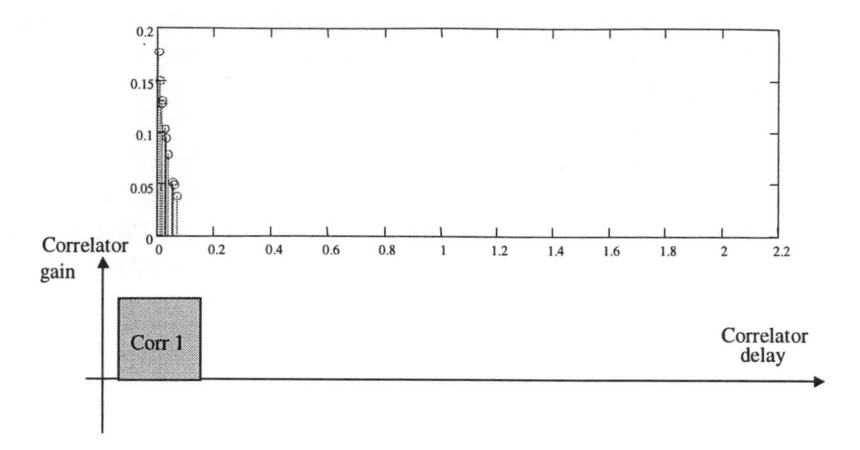

Figure 11.12. An example of indoor channel and of Rake receiver finger delay and amplitude allocation.

11.3 UMTS radio system planning

The main focus in this section is to describe the differences between GSM radio system planning and UMTS radio system planning. Often it is claimed that CDMA planning is much more difficult and complex than planning of TDMA based systems. Partly the claim is true because many aspects of planning are more closely tied to each others in UMTS planning than they are, for example, in GSM planning. However, the principles of good planning are valid for both UMTS and GSM radio interface. Differences exist mainly because of the implementation of the different technologies rather than the planning principles.

11.3.1 UMTS radio system planning process

The UMTS radio system planning process is similar to the GSM planning process. The phases of the planning process are:

- *dimensioning*
- *configuration planning*
- *coverage and capacity planning*
- *code planning (and frequency planning)*
- *parameter planning*
- *optimisation and monitoring.*

Also, the planning targets are the same as they are in GSM. In general the aim is to maximise

- *coverage*
- *capacity*
- *quality.*

The major differences in the radio system planning process in the UMTS happen in the coverage and capacity planning. In the GSM coverage planning was made separately after dimensioning (the base station amount is estimated in the dimensioning phase in the GSM and the base station sites can be planned fairly well without knowing the capacity requirements), and capacity and frequency planning were made in tandem. In the UMTS the coverage and capacity planning have to be made together because of capacity requirement and traffic distribution influence on coverage. The frequency plan can be made separately, or is not needed at all.

All the UMTS radio system planning phases include, of course, the detailed level changes—power budget, antenna line equipment, traffic model, parameter and monitoring changes—compared to the GSM. However, the radio system planning process has to be modified only in the aforementioned part which is depicted in Figure 11.13.

Figure 11.13 UMTS system planning process.

Figure 11.13 shows directly one key issue in the UMTS coverage and capacity planning namely the influence of traffic data that has to be emphasised continuously in the UMTS radio planning. The distribution of the traffic data to the voice and different data calls at each base station coverage area should be known as accurately as possible. Also the location of the different mobile users—or actually the power budget of each mobile user—should be known as exactly as possible. It is of course not possible to know these mobile user locations exactly but the more accurately they can be forecast the better the radio network can be designed, in theory.

The other—not directly seen—key issue in the WCDMA radio coverage and capacity planning is the regional traffic distribution or the existence of traffic hot spots in the radio network coverage area. This issue is related to the base station site locations which should be selected such that they are always placed on the traffic hot spots and as such offering the best power budget for the mobile users served by those base stations. This base station placement significantly reduces power levels in the radio network and thus also interference, and furthermore increases capacity.

In the dimensioning phase a fixed load for all the base stations within the targeted area is assumed. The value for the load can be the maximum acceptable load for the cells or it can be the predicted load during the busy hour. If the highest acceptable load is used the dimensioning is done according to the worst-case scenario and this may lead to an unnecessary high number of sites. Thus it is better to use the predicted load because it will give more realistic results in dimensioning. In the detailed planning phase the traffic distribution is used to allocate the predicted traffic to the planned cells. This may lead to situations in which the load between the cells can vary remarkably—some cells may have a load that is very close to the maximum acceptable load and some cells may have a fairly low load. In the detailed planning phase coverage targets are also checked. In dimensioning it was assumed that traffic is evenly distributed. Also, it was assumed in dimensioning that propagation is similar for all the cells. Thus all the cells are identical in dimensioning. In the detailed planning coverage predictions can be quite different between the cells due to propagation environment and due to traffic distribution.

11.3.2 WCDMA configuration planning

WCDMA coverage planning is commenced from the configuration planning (the power budget calculation). The power budget in the WCDMA as in the GSM takes into account the base station equipment configuration and the base station antenna line configuration. The WCDMA power budget also contains some new parameters that were not used in the GSM power budget. A typical power budget for WCDMA is presented in Figure 11.14.

The power budget is calculated based on the following assumptions:

- uplink bit rate is 64 kbps and downlink bit rate is 144 kbps
- predicted load in uplink is 30 percent and in downlink 50 percent
- 1 W output power at the BTS is reserved for a connection.

General Information	Units		
Frequency (MHz)	MHz	2100	
Chip rate	Mcps	3.84	
Temperature	K	293	
Boltzman's constant	J/K	1.38E-23	
Service information	**Units**	**Urban**	
		UPLINK	**DOWNLINK**
Load	%	30	50
Bit rate	kbps	64.0	144.0
Receiving end	**Units**	**UPLINK**	**DOWNLINK**
Thermal Noise Density	dBm/Hz	-173.93	-173.93
Receiver Noise Figure	dB	3.00	6.00
Receiver Noise Density	dBm/Hz	-170.93	-167.93
Noise Power	dBm	-105.09	-102.09
Interference margin	dB	1.55	3.01
Receiver interference power	dBm	-108.77	-102.09
Total noise (thermal + interference)	dBm	-103.54	-99.08
Processing gain	dB	17.78	14.26
Required E_b/N_0	dB	5.00	4.00
Receiver sensitivity	dBm	-116.32	-109.34
RX antenna gain	dBi	18.00	0.00
Cable loss	dB	4.00	0.00
LNA gain	dB	0.00	0.00
Antenna Diversity Gain	dB	0.00	0.00
Soft Handover Diversity Gain	dB	3.00	3.00
Power Control Head Room	dB	0.00	0.00
Required Signal Power	dBm	-133.32	-112.34
Field Strength	dBμV/m	10.32	31.31
* $Z = 77.2 + 20*\log(freq[MHz])$			
Transmitting end	**Units**	**Uplink**	**Downlink**
TX power per connection	W	0.126	1.00
TX power	dBm	21.00	30.00
Cable Loss	dB	0.00	4.00
TX Antenna Gain	dBi	0.00	18.00
Peak EIRP	dBm	21.00	44.00
Isotropic path loss	**dB**	**154.32**	**156.34**

Figure 11.14. Typical power budget for WCDMA cell.

The presented power budget is divided into five parts, namely

- *General information*
- *Service information*
- *Receiving end*
- *Transmitting end*
- *Isotropic path loss.*

In *General information* the frequency band, chip rate, temperature and Boltzman's constant are given. In *Service information* the bit rates and loads for uplink and downlink are defined. *Receiving end* and *Transmitting end* define the radio links in the uplink and downlink directions. Finally, *Isotropic path loss* defined the maximum achievable path loss between the receiver and transmitter.[10]

11.3.2.1 Receiving end parameters

Thermal noise density defines the noise floor due to the thermal noise. The calculation for the density is given in Equation 11.2. The values used in the calculations are taken from the Figure 11.14 uplink direction.

$$N = kT = 1.38 \cdot 10^{-23} \frac{Ws}{K} \cdot 293K = 4.05 \cdot 10^{-21} W/Hz$$

$$= 10 \cdot \log_{10}\left(\frac{4.05 \cdot 10^{-21} W}{0.001W}\right) dBm/Hz = -173.93\ dBm/Hz$$

Equation 11.2

where

N is thermal noise density
k is Boltzman's constant
T is temperature in Kelvin.

When the *Receiver noise figure* is added to thermal noise density the results are called the *Receiver noise density,* (it defines the noise density at the receiver). *Noise power* defines the noise power at the receiver within the channel bandwidth of the receiver [10]

$$P_N = kTBF = 1.38 \cdot 10^{-23} \frac{Ws}{K} \cdot 293K \cdot 3.84MHz \cdot 1.995$$

$$= 3.099^{-14} W$$

Equation 11.3

$$= 10 \log_{10}\left(\frac{3.099^{-14} W}{1 mW}\right) = -105.09\ dBm$$

where

P_N is noise power
F is noise figure of the receiver
B is chip rate.

In WCDMA there is interference from the original cell and from the neighbouring cells. The degradation of coverage and capacity is modelled with an interference margin. This depends on the load as shown in Equation 11.4.

$$\eta = -10 \cdot \log_{10}(1 - load) \qquad \text{Equation 11.4}$$

where η is interference margin.

The load is always compared to the maximum capacity of the cell (called pole capacity) so the load is always in between 0 percent and 100 percent. Figure 11.15 shows that the interference margin grows towards infinity if load increases to 100 percent.

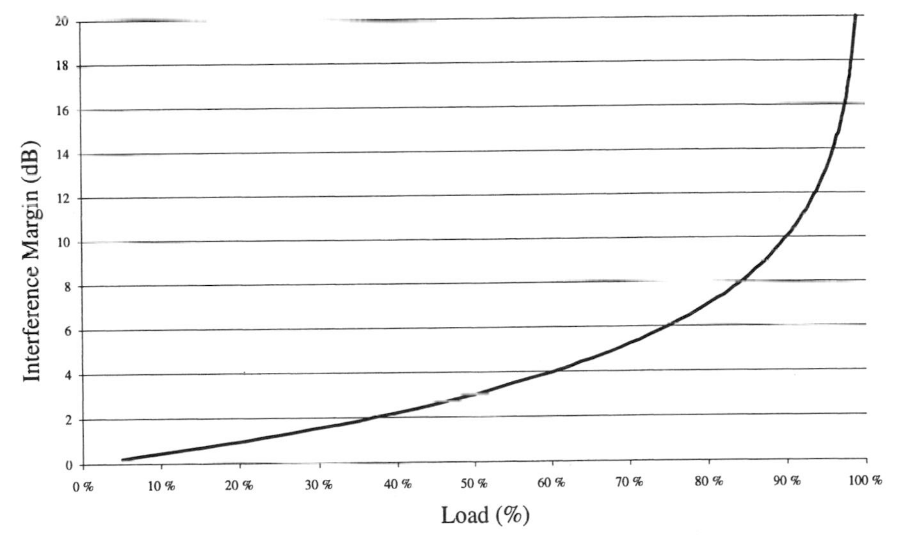

Figure 11.15. Interference margin as a function of load.

Receiver interference power describes the noise level at the receiver due to interference i.e. without thermal noise. The calculation is done by subtracting *Noise power* from the sum of *Noise power* and *Interference margin*. It must be noted that the calculation is done by using the absolute values not dBs. Equation 11.5 shows the calculation.

$$P_{Rx} = (P_N + \eta) - P_N = 10^{(P_N+\eta)/10} - 10^{P_N/10}$$
$$= 10^{(-105.09+1.55)/10}\ mW - 10^{-105.09/10}\ mW$$
$$= 4.42 \cdot 10^{-11}\ mW - 3.99 \cdot 10^{-11}\ mW = 1.328 \cdot 10^{-11}\ mW \qquad \textit{Equation 11.5}$$
$$= 10\log_{10}\left(\frac{1.328 \cdot 10^{-11}\ mW}{1mW}\right) dBm = -108.77 dBm$$

Total noise is the noise floor including thermal noise, noise generated by the receiver (noise figure) and the interference.

Processing gain is a gain that is achieved due to spreading. The gain can be calculated with Equation 11.6.

$$PG = 10\log_{10}\left(\frac{B}{W}\right) = 10\log_{10}\left(\frac{3.84 Mcps}{64\ kcps}\right) = 17.78 dB \qquad \textit{Equation 11.6}$$

where
- PG is processing gain,
- B is chip rate and
- W is the bit rate of the information (symbol rate).

Required E_b/N_0 that is needed to be able to demodulate the signal shows ratio between the received energy per bit and noise energy. The value should be selected based on the service. Usually equipment manufactures indicate E_b/N_0 values for different services that can be used for their equipment.

Receiver sensitivity is the signal level that is required at the antenna port of the receiver to be able to achieve acceptable quality in receiving. This value is comparable to the sensitivity that is presented in a link budget for GSM (recall Chapter 3).

The antenna line parameters for a WCDMA cell are very similar to the parameters for a GSM cell. There are some remarks that must be made:

- LNA will be used in almost all cells. It improves uplink coverage but also uplink capacity.
- Soft handover gain is not used in GSM. In UMTS a mobile can communicate with several base stations at the same time. This improves the performance of the receiving end. However, sometimes soft handover gain is included in E_b/N_0

and in this case soft handover gain should not be included in link budget as a separate item.

- Power control headroom is a parameter that is used to reserve some power for the situations where power control cannot compensate deep dips (i.e. fast fading). This occurs usually on the cell edge. The power control headroom reduces cell range when reserving some power for compensating fast fading.

11.3.2.2 Transmitting end

Transmitting end in the WCDMA power budget is similar to GSM power budget. There are some differences as listed below.

- BTS power is given per user in WCDMA. In GSM one user gets the full power of a TRX when using the time slot. In WCDMA the output power of a BTS is shared between the control channel and all the users in a cell. In power budget calculations usually the maximum power per connection is defined to model the downlink direction in a realistic manner. If a user could use the full power of a WCDMA BTS this would mean that there cannot be other users in that cell at that moment.
- In WCDMA base stations there are not similar combiner that are used in GSM thus there are no combiner loss in the power budget.

11.3.2.3 Isotropic path loss

Isotropic path loss is calculated in the same way as it is for a GSM cell. In WCDMA power budget there can be remarkable differences in isotropic path losses between uplink and downlink. The main reasons for these differences are asymmetry in traffic and different bit rates in the uplink and downlink. Asymmetry in traffic is expected to happen especially in packet data. The ratio can be as high as 1:10 (UL:DL) or even higher. Partly this is reflected to the bit rates in the uplink and downlink directions.

The isotropic path loss is used for cell range calculations in a similar way as it is used in GSM. In Figure 11.16 cell range for the urban area is calculated based on the isotropic path loss, antenna heights and slow fading margin.

Cell sizes	Units	Urban	
MS antenna	m	1.5	
BS antenna	m	25.0	
Standard	dB	7.0	
BPL Average	dB	15.0	
BPL	dB	8.0	
Okumura-Hata (OH)	**Units**	**Urban**	
Area Type Correction	dB	-2.0	
Indoor coverage	**Units**	**Urban**	
Propagation Model	OH/WI	OH	
Location Probability over Cell	%	95.0	
Slow Fading Margin +	dB	23.8	
		UPLINK	**DOWNLINK**
Coverage	dBμV/m	34.1	55.1
Coverage	dBm	-109.5	-88.5
Cell Range	km	0.64	0.72
Outdoor coverage	**Units**	**Urban**	
Propagation Model	OH/WI	OH	
Location Probability over Cell	%	95.0	
Slow Fading	dB	7.4	
		UPLINK	**DOWNLINK**
Coverage	dBμV/m	17.7	38.7
Coverage	dBm	-126.0	-105.0
Cell Range	km	1.83	2.09
Cell range		**Indoor Uplink**	**0.64 km**

Figure 11.16 Cell range calculation for an urban area. Isotropic path loss is taken from Figure 11.14.

11.3.3 WCDMA coverage and capacity planning

Coverage and capacity planning in WCDMA are tied closely together. In low traffic areas planning of WCDMA is quite similar to GSM because load does not have great impact on coverage. Of course, there are many details that differ between the systems but the main principles can be applied to both systems. In high traffic areas planning of GSM is focusing on frequency planning—the better frequency plan that can be achieved the higher capacity is obtained. Also, cell splitting in GSM can be considered as a straight forward process. In WCDMA there is no such clear split between coverage, interference and capacity.

11.3.4 WCDMA propagation prediction and coverage planning

The propagation predictions contain in the WCDMA the same planning phases as in the GSM. First, the base station configuration and thus the power budget have to be defined. At the same time the dimensioning exercise has to be done in order to decide the suitable average effective antenna height over the service area. It has to be remembered that the antenna height definition has as significant an influence on the radio network infrastructure as on the base station site locations or site distances. Also the coverage threshold has to be well defined in order to exceed the required quality criteria but not to make extra investments for the radio network elements. Moreover the capacity targets and forecasts has to be well known at this phase because they have a strong effect on the base station coverage areas.

When the base station antenna height, coverage threshold and the capacity requirements are defined and the base station configuration is clarified in the power budget calculations the actual propagation prediction process can start. This process begins with the propagation measurements also called radio propagation verification measurements, for tuning the prediction model(s). These measurements are used and needed especially for the prediction models for the design of macro base stations. In the micro or pico base station environments the measurements are also required but they are not used as much for the prediction model verifications (but the measurement results are used without prediction models). After conducting the measurement campaign over the WCDMA service area the prediction model has to be tuned. When the prediction model is tuned the final base station parameters can be used to make the propagation predictions.

11.3.5 WCDMA planning margins and coverage thresholds

The final coverage planning starts when the coverage prediction models are tuned and the real base station and base station antenna line configuration parameters are added to the coverage planning tool. The evaluation of the optimised base station locations can be done when the planning criteria—planning threshold is defined. This planning threshold means that the reasonable QOS level for the different geographical locations have to be agreed: first major national areas, cities and roads where coverage has to exist and then sub-areas of them such as urban and suburban areas. The planning threshold also concerns whether the service has to be extended inside vehicles and buildings in different areas.

The planning threshold itself is defined as in the GSM by starting from the mobile station sensitivity (threshold is for the downlink direction) and by adding the required planning margins to the sensitivity value. The required margins are slightly different in the WCDMA and they are

- slow fading margin (shadowing)
- macro diversity or soft handover gain
- power control head room
- body loss, antenna orientation loss
- in-vehicle or indoor penetration loss
- interference margin.

The planning threshold is calculated by adding all these components to the mobile station sensitivity as shown in Equation 11.7.

$$planning\_threshold = MS\_sensitivity + slow\_fading\_margin - macrodiversity\_gain + power\_control\_head\_room + body\_loss + orientation\_loss + penetration\_loss + \eta \qquad \textit{Equation 11.7}$$

In Equation 11.7 it should noted that sometimes macro diversity gain is included in sensitivity. In that case *macrodiversity_gain* should not be used in calculating the planning threshold.

11.3.6 WCDMA capacity planning

WCDMA capacity planning is directly related to the power budget and thus to the base station coverage area. In the power budget in Figure 11.14 only one type of service (64/144 kbit/s data transmission) was introduced and thus the base station coverage area is fixed for this service. There is a possibility of having any type of service between the voice calls and 2 Mbit/s data traffic in the WCDMA base station.

This means that the base station coverage area (a coverage boundary) is different for different users in the downlink and uplink directions (see Figure 11.17). The coverage area is reduced in the downlink because of the increased power requirement for the connection: the minimum energy per bit at the reception is constant but the bit rate increases. Correspondingly in the uplink direction the interference increases—or the base station coverage area reduces—if the transmission rate is larger.

Basically the question is about the processing gain, PG, which varies significantly when comparing the 12.2 kbps voice call (PG = 25 dB) and 2 Mbps data transmission (PG = 2.8 dB) connections.

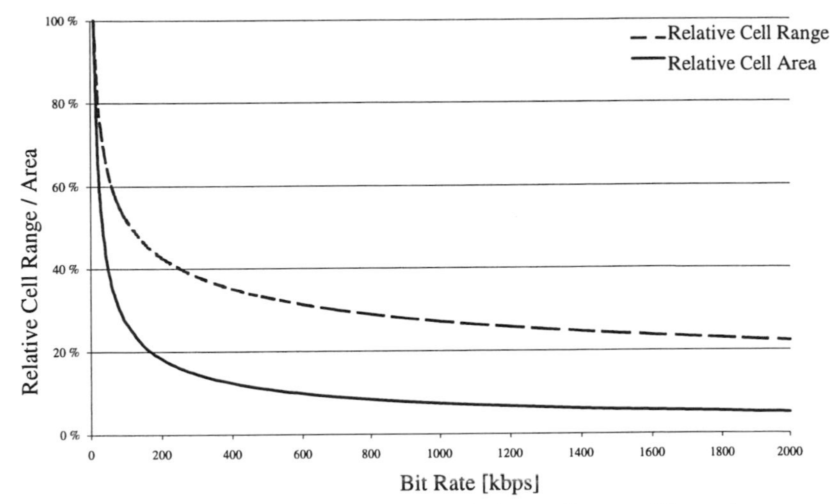

Figure 11.17. Relative cell range and relative cell area. The cell ranges are calculated by using Okumura-Hata propagation formula and antenna height of 25 meters.

In the uplink direction the main objective in capacity planning is to limit interference from the other cells to an acceptable level. Equation 11.8 gives the load in the uplink direction.

$$\eta_{\text{uplink}} = \sum_{n=1}^{N} \frac{1}{1 + \frac{W/R_n}{(E_b/N_0)_n \ \alpha_n}} (1+i) \qquad \textit{Equation 11.8}$$

where

- i is interference from other cells
- N is total number of active users in a cell
- W is bandwidth
- R_n is user bit rate
- E_b is energy of bit
- N_0 is energy of interference
- α_n is activity (e.g. voice activity).

Network planning can affect the uplink load by reducing i. This is done by avoiding unnecessary overlapping of the cells. This can achieved by using buildings, hills, etc., as obstacles blocking the interfering cells. Also, tilting will be a very useful tool in limiting the interference.

In the downlink directions (Equation 11.9) there are two aspects that should be considered. First, the interference from the other cells, *i*, and the power of the base station. The load equation for the downlink is similar to the equation for uplink. However, in the downlink equation there is a new parameter called *orthogonality*, v. Orthogonality shows the degree in which the users in the same cell interfere with each other. If orthogonality is 100 percent the users do not cause interference for other users in the same cell. [10]

$$\eta_{\text{downlink}} = \sum_{n=1}^{N} \alpha_n \frac{(E_b/N_0)_n}{W/R_n} [(1-v)+i] \qquad \textit{Equation 11.9}$$

Because the output power of the base station transmitter is shared by all the users in the cell it is possible that the power of the transmitters limits the capacity in downlink direction. The required power (P) at the base station is given in Equation 11.10.[10]

$$P_{BTS} = \frac{P_{noise}^{total} W \bar{L} \sum_{n=1}^{N} \alpha_n \frac{(E_b/N_0)_n}{W/R_n}}{1-\eta_{\text{downlink}}} \qquad \textit{Equation 11.10}$$

In Equation 11.10 L is an average path loss between the base station and a mobile.

In the downlink direction network planning can affect interference from other cells, *i*, and the average loss between base station and mobile, *L*. Interference from other cells can be reduced by designing dominance areas of the cells as well as possible. The path loss between the base station and the mobiles can be minimised, for example, by selecting the right antenna locations. Also, repeaters and indoor solutions will increase capacity by reducing the path loss.[10]

11.3.7 WCDMA code and frequency planning

In the WCDMA code and frequency planning are seen as a simple task from a network planning point of view. The system takes care of most of the code allocation. The main task for network planning will be the allocation of scrambling codes for the downlink. There are 512 set of scrambling codes available so it can be said that the code reuse for downlink is 512. This means that code allocation is a pretty simple task. However, it is recommended that the allocation is done with the help of a planning system because there is always possibility for an error if the allocation is done manually.

Frequency planning will have minor importance compared with GSM. At most the UMTS operators have two or three carriers thus there is not much to plan. However, there are a few decisions that the operators have to make,

- which carrier(s) is used for macro cells?
- which carrier(s) is used for micro cells?
- any carrier(s) is reserved for indoor solutions?

When making the decisions the interference aspects should be considered. The selection of the carriers may have an impact on intra-operator and inter-operator interference. For example, micro cells can locally cause quite high interference for operator's macro cells or another operator's macro or micro cells. Many of the potential problems can be solved by proper network planning and one of the tools to solve these problems is the selection of frequencies.

11.3.8 WCDMA optimisation and monitoring

The WCDMA like the GSM system needs continuous monitoring because the mobile users' location and traffic behaviour varies all the time. This monitoring requirement is only emphasised in the WCDMA because the traffic demand can vary strongly and this variation influences directly the radio network quality. The better and more accurately the traffic amount and locations can be modelled the better and more efficiently (cost, quality, etc.) the radio network can be designed and implemented.

The indicators that should be monitored are, for example,

- traffic
- traffic deviation
- traffic mix
- soft handover percentage
- average TX power
- average RX power
- drop calls
- handovers per call
- handovers per cell
- inter-system handovers
- throughput
- BER, BLER, FER

Many of the listed indicators should be collected on a cell and service basis because it may give some hints on how to optimise the parameters to enhance the performance of the network.

11.4 Conclusions

The UMTS radio interface system planning has the same basic philosophy as the GSM but varies in the detail mainly because of the change of radio propagation channel that is a wideband type. The major subjects and findings are again gathered, in Table 11.2, to summarise the major issues concerning radio interface system planning in the UMTS.

Table 11.2 UMTS radio interface system planning

Subject	**Findings**
UMTS key instruments	–Data –Service –Implementation
WCDMA radio propagation channel	–RAKE receiver takes into account delay spread –Delay spread versus diversity reception? –The fast fading component still exists –The slow fading is constant
WCDMA coverage and capacity planning process	–Traffic information and forecast –Traffic hot spots
WCDMA power budget	–Load factor is typically 0.5
Coverage threshold	–As in the GSM –Power control margin –Soft handover (macrodiversity) gain
Capacity planning	–Processing gain PG depends on the service
Frequency planning	–Not needed
Optimization and monitoring	–Has great importance

11.5 References

[1] 3GPP Technical Specification 25.213, Spreading and Modulation (FDD).

[2] 3GPP Technical Specification 25.223, Spreading and Modulation (TDD).

[3] 3GPP Technical Specification 25.104, UTRA (BS) FDD; Radio Transmission and Reception.

[4] 3GPP Technical Specification 25.101, UE Radio Transmission and Reception (FDD).

[5] T. Ojanperä, and R. Prasad, "Wideband CDMA for the Third Generation Mobile Communications," Artech House Publishers, 1999, p. 439.

[6] 3GPP Technical Specification 25.212, Multiplexing and Channel Coding (FDD).

[7] 3GPP Technical Specification 25.211, Physical Channels and Mapping of Transport Channels onto Physical Channels (FDD).

[8] 3GPP Technical Report 25.943, Deployment Aspects.

[9] W.C. Jakes, Jr., (ed.), "Microwave Mobile Communications," Wiley-Interscience, 1974.

[10] A. Toskala, and H. Holma, "WCDMA for the UMTS," Wiley-Interscience, 2000.

INDEX

RECEIVED
MAY 17 2002